Rainfall-Induced Landslides Hazard

Rainfall-Induced Landslides Hazard

Editor

Clemente Irigaray

MDPI • Basel • Beijing • Wuhan • Barcelona • Belgrade • Manchester • Tokyo • Cluj • Tianjin

Editor
Clemente Irigaray
Universidad de Granada
Spain

Editorial Office
MDPI
St. Alban-Anlage 66
4052 Basel, Switzerland

This is a reprint of articles from the Special Issue published online in the open access journal *Hydrology* (ISSN 2306-5338) (available at: https://www.mdpi.com/journal/hydrology/special_issues/Rainfall_Landslides).

For citation purposes, cite each article independently as indicated on the article page online and as indicated below:

LastName, A.A.; LastName, B.B.; LastName, C.C. Article Title. *Journal Name* **Year**, *Volume Number*, Page Range.

ISBN 978-3-0365-2177-0 (Hbk)
ISBN 978-3-0365-2178-7 (PDF)

Cover image courtesy of Clemente Irigaray

Contents

About the Editor

Clemente Irigaray (Prof.) is a Full Professor of Terrain Engineering in the Department of Civil Engineering at the University of Granada (UGR), Spain. He is a founding member and current principal investigator of the Research Group RNM-121 "Environmental Research, Geological Risks and Land Engineering" of the Regional Government of Andalusia. He has co-authored more than 200 technical journal and conference papers and has served as an editor and reviewer of several journals and congress proceedings. He has made several stays in international teaching centers and currently participates as a visiting professor at the UTPL (Ecuador). His research activity has been focused on the development of new remote sensing and GIS techniques for geological hazard forecast, monitoring and mitigation, especially in landslide dynamics, mechanisms and processes as well as numerical and empirical methods for landslide-hazard assessment.

Preface to "Rainfall-Induced Landslides Hazard"

Landslides can cause human injury, loss of life and economic devastation, and destroy construction works and cultural and natural heritage. The level of risk as a consequence of landslides depends on the hazard associated, that is, on the recurrence or temporal probability that one or more landslide process will affect a terrain unit. This is why knowledge into the frequency with which these processes occur or important activity changes arise, together with the different types of landslides and their sizes or intensity classes, provides valuable information to aid in predicting the transient behavior linked to the destructiveness of these natural events. Landslides are frequently triggered by hydro-meteorological phenomena, mainly as a consequence of intensive rainfall, and have led to losses of millions of dollars and thousands of fatalities. In recent decades, there has been significant improvement in landslide observation with hazard assessment modeling using numerical and analytical methods. These developments offer new displays for modeling landslide hazards, leading to new insights into their functioning and new approaches to process modeling to obtain better prediction. Within this framework, the challenge of this book is to describe the latest developments and applications of these numerical and analytical methods to improve our understanding of rainfall-induced landslide models and other aspects of landslide hazard. With this aim, a discussion on this topic is stimulated, collecting a number of manuscripts recently published in a Special Issue from the journal Hydrology, which focused on the benefits obtained by the use of new algorithms and measurement systems for landslide-hazard assessment.

Clemente Irigaray
Editor

 *hydrology* 

Article

Rainfall-Induced Landslides and Erosion Processes in the Road Network of the Jaén Province (Southern Spain)

Ramón Carpena [1,2], Joaquín Tovar-Pescador [3,4], Mario Sánchez-Gómez [4,5], Julio Calero [4,5], Israel Mellado [5], Francisco Moya [6] and Tomás Fernández [4,6,*]

1 Department of Mechanical Engineering, University of Jaén, 23071 Jaén, Spain; rcarpena@ujaen.es
2 Municipal Infrastructures Area, Deputy of Jaén, 23071 Jaén, Spain
3 Department of Physics, University of Jaén, 23071 Jaén, Spain; jtovar@ujaen.es
4 Centre for Advanced Studies in Earth Sciences, Energy and Environment, University of Jaén, 23071 Jaén, Spain; msgomez@ujaen.es (M.S.-G.); jcalero@ujaen.es (J.C.)
5 Department of Geology, University of Jaén, 23071 Jaén, Spain; jmellado@ujaen.es
6 Department of Cartographic, Geodetic and Photogrammetric Engineering, University of Jaén, 23071 Jaén, Spain; fmoya@ujaen.es
* Correspondence: tfernan@ujaen.es; Tel.: +34-958-212843

Abstract: Rainfall thresholds are one of the most widely applied methods for indirectly estimating landslide return periods, which are subsequently used in hazard analyses. In this study, the starting point is an incidence database of landslides and erosive processes affecting the road network of the province of Jaén (southern Spain), in which the positions and dates of civil repair works can be found. Meanwhile, the use of a daily rainfall database in a dense grid (1 km) allowed for the estimation of the rainfall series at each incidence point with high precision. Considering the news in the local media and applying spatial proximity, temporal proximity, and maximum return period criteria, rainfall events of various duration (1 to 90 days) could be associated approximately with each point. Then, the rainfall thresholds and their return periods were estimated. A linear equation was adjusted for the rainfall duration threshold (E = 6.408 D + 74.829), and a power-law curve was adjusted for the intensity–duration pair ($I = 47.961\ D^{-0.458}$). Non-significant differences were observed between the thresholds and the return periods for the lower and higher magnitude incidences, but the durations for the former were lower (1–13 days), compared to those of the latter (7–22 days). From the equations, rainfall events of different durations could be estimated for use in hazard analysis, as well as for the future development of warning systems.

Keywords: rainfall-induced; landslides; erosion processes; road network; Jaén province; rainfall thresholds

Citation: Carpena, R.; Tovar-Pescador, J.; Sánchez-Gómez, M.; Calero, J.; Mellado, I.; Moya, F.; Fernández, T. Rainfall-Induced Landslides and Erosion Processes in the Road Network of the Jaén Province (Southern Spain). *Hydrology* **2021**, *8*, 100. https://doi.org/10.3390/hydrology8030100

Academic Editor: Luca Brocca

Received: 31 May 2021
Accepted: 1 July 2021
Published: 5 July 2021

1. Introduction

Landslides are considered one of the most important natural hazards worldwide, causing thousands of victims per year and costs worth billions of euros [1–6]. Landslides originate in different mountainous regions in Europe [7], such as the Alps, Norway, and the Mediterranean countries [8,9], and specifically in Spain and the Betic Cordilleras [10–12]. Compared with other risk phenomena, such as earthquakes or floods, the effects of landslides are more diffuse and continuous in space and time; thus, according to some studies, their impact has been underestimated [4]. Despite this, they produce significant damage to infrastructure, properties, and the environment itself, as well as interrupting socio-economic activity [1–3].

One of the most effective measures for risk prevention and mitigation is its evaluation which, according to the classic formulation of Varnes [1], includes both the hazard (probability that a potentially harmful phenomenon occurs in a given space and time) and the exposure and vulnerability of the elements at risk. For the former, there exist deterministic methods based on precise knowledge of the factors conditioning the instability [13]. These include geotechnical properties, terrain morphology, and the hydrological conditions usually conditioned by rainfall as the triggering factor [14–18]. However, the difficulty of obtaining accurate data [19], especially for studies carried out in a more or less extensive area, leads to probabilistic methods being more frequently applied in hazard studies. Probabilistic methods are based on correlation analyses between determinant factors and landslides, both by means of bivariate approaches [12,20,21] and multivariate ones [3,21–23]; however, in recent years, machine learning techniques have become increasingly used [21,24]. The first step in these analyses is to elaborate inventories or databases that collect the spatial locations, occurrence times (dating), and thematic attributes of the movements [11,25]. For this, direct observation, geomatics data capture techniques, and in situ or laboratory tests of the different terrain properties are used [11,26].

Among them, dating is usually one of the most complicated issues for landslides, due to their aforementioned diffuse and continuous nature [27,28]. Direct dating requires recording by direct observation or in-situ sensors; however, geomatics techniques, such as Global Navigation and Satellite Systems (GNSS), photogrammetry, optical remote sensing (ORS), LiDAR, and Interferometry of Synthetic Aperture Radar (InSAR), have allowed for important advances, especially following their spatial and temporal resolution increases [29,30]. Another option is indirect dating from triggering factors [26,31], such as earthquakes and/or rainfall, which are more easily recorded by different instruments, usually gauges or digital sensors. Considering that rainfall is the triggering factor in most cases, it is necessary to establish the relationship between rainfall and landslides, which have been done in numerous studies worldwide [32–38].

Many of these studies have allowed the calculation of rainfall thresholds using empirical methods [9,19,28,36,39–54], especially for shallow movements (debris and mud flows, shallow slides and even erosion processes, especially gullies [55,56]), where the correlation with intense rainfall events is clearer, although it has also been applied to deep ones [40]. As in the aforementioned deterministic methods [14–18], these thresholds are based on the fact that an increase in rainfall leads to a change in the hydrological soil conditions and then in slope instability. Thus, some approaches have used not only the triggering rainfall, but the antecedent rainfall that influences the hydrological conditions [16,18,19,32,35,40,57,58] and, in some cases, so-called hydro-meteorological thresholds have been developed [59–62]. Other studies have considered the influence of conditioning factors on the thresholds [45,50], even the main annual precipitation (MAP) [47,63].

In some works, very precise knowledge of the rainfall data (hourly), as well as the moment in which the movement starts, have been used [9,19,41–50,53]; meanwhile, in others, only the daily rainfall data are known [28,36,40,51,52]. There are even cases in which the landslide time or date can only be approximated [36,40,43,51] and reconstructed after subsequent inventories and/or reviews from the news found in the media [28,40,51,52]. In any case, following [42], different variables can be used to define these thresholds, such as the total event rainfall (E), rainfall event–duration (E–D), intensity–duration (I–D), and rainfall event–intensity (E–I). Although most studies have been carried out at the local level, there are some cases of application over large regions [42–46], as well as works and databases that collect indices all over the world [43,54]. Moreover, some works have led to the development of algorithms and computational tools that can calculate these thresholds [48,49].

Once the thresholds are determined, they can be applied to predict the probability of landslide occurrence by determining the return periods [19,28,36,40,41,51,52], which can be incorporated into the corresponding hazard maps [64,65]. Likewise, they can be integrated into (early) warning systems [21,63,64], which can prevent the population and authorities from being subjected to landslides in those cases in which the rainfall threshold is reached.

The objective of this study is the determination of rainfall thresholds that cause landslides or erosion processes associated with the road network of the Jaén province. For this, an incidence inventory or database between 1997 and 2013 and a rainfall database between 1971 and 2016 are available. Then, news from the local media were also considered in the analysis that allowed associating incidences with rainfall events. This has led to the calculation of rainfall thresholds and the return periods, thus allowing for hazard modelling in the province of Jaén and the development of warning systems in future works.

2. Materials and Methods

2.1. Study Area

The study area (Figure 1) corresponds to the province of Jaén (13,486 km^2). It coincides approximately with the natural region of the eastern or upper Guadalquivir River basin. The altitude ranges between 152 and 2160 m, with an average value of 715 m. The average slope is 12.21°, although it is also very variable between the mountain ranges and the lower lands of the Guadalquivir valley.

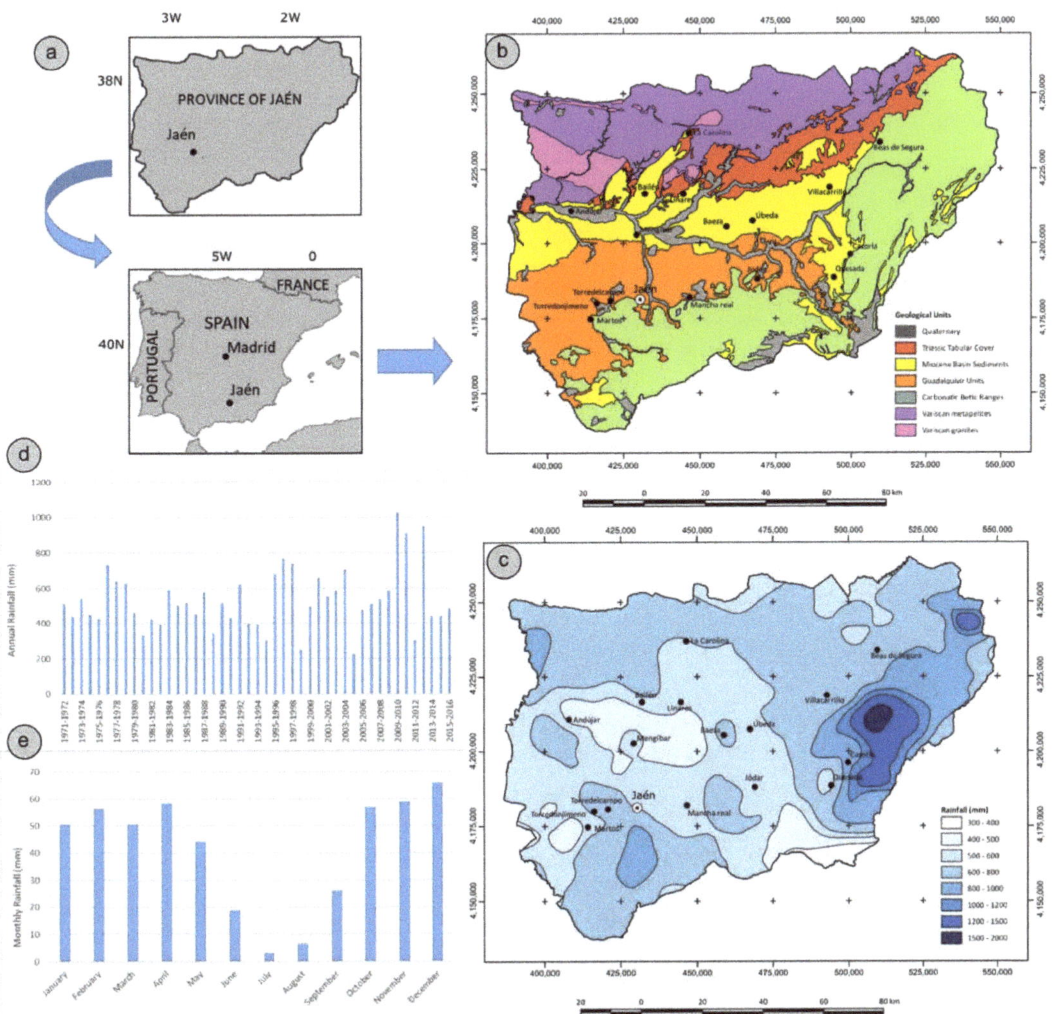

Figure 1. Location, geological setting, and climatic characterization: (**a**) location; (**b**) geological setting; (**c**) mean annual precipitation (MAP) of the province; (**d**) series of annual precipitation; and (**e**) monthly precipitation.

From the geological and morphological point of view, three domains can be distinguished which, from South to North, are as follows (Figure 1b):

- The External Zones of the Betic Cordillera, which are made up of mesozoic and cenozoic carbonate or loamy-clayey rocks, structured as a fold and thrust belt from the lower Miocene to the present [66], in which several paleogeographic domains appear (Prebetic and Subbetic). The Betic External Zone form several mountain ranges (Sierra Cazorla and Segura, Sierra Mágina and Sierra Sur of Jaén), partially isolated by main rivers and tributaries of the Guadalquivir River.
- The sedimentary infill of the Guadalquivir basin, differentiated into two parts. In the north, the Guadalquivir basin is filled with Miocene loamy and clayey sediments, which are slightly deformed and which overlie the tabular cover of the Iberian Massif, made up of Triassic clays and sandstones and Jurassic limestones. In the south, the infill of the basin is highly deformed by the Betic Miocene displacements, which incorporate tectonically Betic soft materials as Triassic evaporites (salt and gypsum) or Cretaceous clayey marls [66].
- The Variscan Domain, which constitutes the outcropping basement of the Iberian Massif, in which metapelites (slates, grauwackes, and so on) and intruding igneous rocks (granites and granodiorites) are the predominant lithologies.

Over all these materials, quaternary deposits related to present fluvial dynamics and slope sediments are located.

From the climatic point of view, the province of Jaén corresponds mostly to the hot summer Mediterranean (Csa de Koppen) [67] climate type. More specifically, it can be catalogued as the Mediterranean meridional type of the Guadalquivir valley [68]. This is characterized by a mean annual precipitation (MAP) between 500 and 650 mm, with maximum values distributed between the autumn, winter, and spring and minimum values in the summer. Meanwhile, the average temperatures are 17–18.5 °C, with very pronounced maximum values in summer. However, there are sectors in the province with MAP higher than 1000 mm in the mountain ranges (Mediterranean mountain), and others with MAP that does not reach 400 mm in the southeast (Mediterranean arid), as shown in Figure 1c.

Thus, at a central point representative of the average physical conditions of the province (point 066 of the incidence database, see below), the MAP was 533 mm within the period considered (1971–2016), with a minimum value of 223 mm in the hydrological year 2004–2005 and a maximum value of 1026 mm, which was reached in 2009–2010 (Figure 1d). This wide interval shows the variability of precipitation over the years, with a standard deviation of 172 mm and coefficient of variation of 0.32. Within the year, rainfall was higher between November and April (50–60 mm) and lower between June and September (below 25 mm), with a monthly average rainfall of 41.4 mm (Figure 1e).

The predominant land-use is agricultural crops; within them, olive grove constitutes 44% (5928 km^2) of the province's surface area [69]. To a lesser extent, other crops (e.g., cereal) and areas of natural vegetation appear in the mountains, along with scrub and coniferous/hardwood forests.

The province has a population of 638,000 inhabitants, with only 2 urban areas exceeding 50,000 inhabitants [70], and a low industrial activity focused on agriculture. On the other hand, it has a road network of different orders (state, regional, and provincial), in which the A-44 and A-316 highways stand out. This study is focused on the extensive and penetrative road network of the Provincial Deputy of Jaén, which is about 1600 km long. This network is a fairly representative sample of the different physical environments of the province and, so, its study can provide valuable information on the instability conditions, not only in the network environment, but also in the whole province.

2.2. Incidence Database

The overall methodology followed in this study is shown in Figure 2. The first step is the elaboration of the database of incidences on the road network of the Jaén province.

This includes data extracted from the files of the works carried out for their repair and maintenance, completed with field data and other data extracted from previous maps. The database has been elaborated through the years, by gathering information on road interventions that took place from 1998 to 2013.

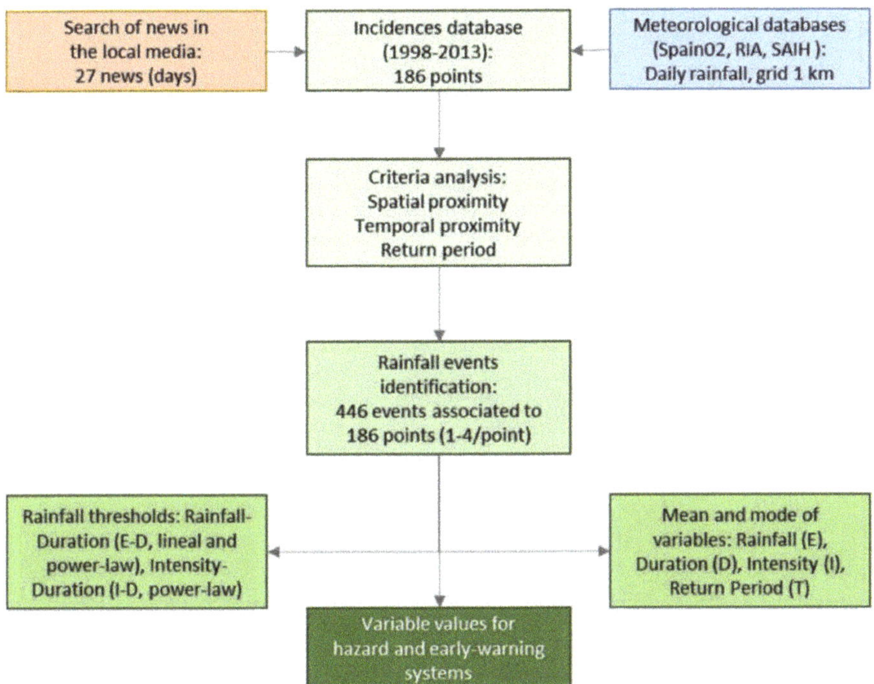

Figure 2. Methodology workflow.

Thus, the original database included data identifying the incidence (coordinates, project, works, and so on); vegetation and land-use; geomorphological and topographic data; geotechnical data (load capacity, constructive conditions, and so on); hydrogeology (drainage and permeability); description of the incidence (year, month, typology of incidence, roads, kilometres, and so on); geology (lithology and surface formations); road data (road surface, slope, curvature radius, and so on); and, finally, the constructive solution adopted. The database was subsequently tested on the ground, especially with regard to geology and the descriptions of the incidences, as well as morphological aspects.

Finally, the recorded and reviewed incidences were digitized onto the orthophotography and subsequently refined using several GIS tools (e.g., clipping, buffering). The result was an enriched incidence inventory or database of the road network of the province. A basic distinction was made between two categories: the lower magnitude and shallower processes that affect the road surface, road cuts, and embankments; and the higher magnitude processes that involve a certain general slope instability. Among the former, the following types were differentiated: erosive processes (gullies), undercut of road embankments, and small slides and collapses of the road cuts. Among the latter, slides, earth or mud flows, and creeping processes were differentiated. The inventory is presented and described in Section 3.1.

2.3. Rainfall Data Processing

Different meteorological databases were used to estimate the rainfall series:

- Spain02 Database, high-resolution daily precipitation data, developed by the Institute of Physics of Cantabria (Spain) and the Spanish Meteorological Agency (AEMET) from a dense network of more than 2500 quality-controlled stations for precipitation and near 250 for temperatures. The Spain02.v5 provides daily data from 1951 to 2015, gridded in increments of 0.1°, corresponding approximately to a resolution of 10 km [71,72].
- RIA database, a network of agroclimate information by the Department of Agriculture, Fisheries, and Rural Development of the Andalusian Government [73]. It contains updated data on the networks of automatic meteorological stations (~120 stations), which are equipped with electronic sensors and distributed throughout the Andalusian territory.
- Database of the network of the Automatic Hydrological Information System (SAIH) from the Authority of the Guadalquivir River Hydrographical Basin [74].
- Data obtained by the Atmosphere and Solar Radiation Modeling (MATRAS) research group from the Weather Research and Forecasting (WRF) model using the Integrated Forecasting System (IFS) reanalysis data provided by the European Centre for Medium-Range Weather Forecasts (ECMWF) and local data of the meteorological station of the University of Jaén [75].

The data processing comprised the integration of the previous databases and the application of physical and statistical filters, which allowed us to obtain an interpolated regular grid at 1 km. This grid was used to assign a daily rainfall value, from 1971 to 2016, to each of the 186 incidence points of the road network in the province of Jaén. The value assigned was that of the closest grid node which, taking into account its resolution, is a point located at a small distance (lower than 1 km). For the daily precipitation values in the province of Jáen, three zones can be considered, in two of which the WRF underestimates the precipitation and, in the other, it is weakly overestimated. To reduce the uncertainty of the data provided by the WRF, we used data obtained at ground stations as control points. Using geostatistical techniques, the uncertainty was less than 12% in all cases. Therefore, although the densification of the grid tended to smooth the real values, the estimation error and the uncertainty derived was low.

2.4. Rainfall Event Identification

The identification of rainy events associated with the landslides and erosive processes in the study area was based on relating the incidence database with the rainfall series.

First, from the daily data, the accumulated rainfall over 2 days, 3 days, 5 days, 7 days (1 week), 10 days, 15 days, 30 days (1 month), 45 days, 60 days (2 months), 75 days, and 90 days (3 months) were calculated, in order to analyse the influence of short- and medium-term rainfall on the generation of landslide and erosion processes.

Then, different rainfall variables could be defined: rainfall amount associated with the event (E) in mm; duration of the event (D) in days; and the intensity (I), which was calculated as the relationship between the rainfall and the duration of the event and expressed in mm/day. For each rainfall event–duration (E–D) pair, the probability of exceedance and return period (T) in years were calculated considering a Weibull series, as in previous studies [36,40,51,52].

Meanwhile, the incidence database only included information about the month in which the civil work started, while the accurate date when the incidence occurred remained unknown. Thus, additional information was used, in order to estimate a more precise date. For this, we used information found in the media, especially the news published in the local and regional press, such as the IDEAL newspaper [76], which has a historical record since 2006 and can be accessed freely on the internet. This approach has been used in some previous works [28,40]. After a deep search, based on terms such as landslides (and their synonymous terms in Spanish), road affected, traffic interruption, and so on, a total of 98 news items were found between 2006 and 2013, of which 27 events were directly related to the incidences (i.e., landslides or erosive processes). These are summarized in Table 1.

Table 1. News about incidences registered in the local media: S, slides; R, (rock)falls; M, mud and earth; U, road undercut; W, water, F, floods; T, traffic cuts. Sectors: SS, Sierra Sur; SM, Sierra Mágina; SC, Sierra Cazorla; SG, Sierra Segura; J, Jaén (capital city); CH, Central Hills; W, Western sector; N, Northern sector; Gen., general.

Date	Description	Zone
12/09/2006	R, M, W, F	SS, W
06/04/2007	R, M	SG
10/09/2008	M, W, T	CH, N
08/08/2009	R	N
11/08/2009	M, T	SS
25/12/2009	M, T	Gen.
28/12/2009	R, M, F, T	SS, SC, J, CH
07/01/2010	W, T	SS, SM, SC, SG, W
11/01/2010	R, U, F, T	SS, SC, W
15/01/2010	S, R	SS, W
19/01/2010	S, R, U, T	SS, SC, J, CH
19/02/2010	S, R, U	Gen.
21/02/2010	S, R, U	SS, SC, SG, J, CH
23/02/2010	R, F, T	Gen.
07/03/2010	S, T	SS, J, W
10/03/2010	S, U, F	SS, SM, CH
30/10/2010	R, W, T	SM, CH
08/12/2010	R	SS, J, N
20/12/2010	R, M, W	Gen.
02/05/2011	M, T	CH, N
04/11/2012	R, W, T	Gen.
06/11/2012	T	J, W, N
08/11/2012	T	SC, CH, W, N
11/03/2013	S, M, U, T	SG, CH, W, N
13/03/2013	S, M, U, T	Gen.
19/03/2013	R	SS
01/04/2013	R, F, T	SC, SG, CH, W, N

Then, three criteria were applied: the spatial proximity, the temporal proximity, and the magnitude of the rainy event.

1. The spatial proximity between the approximate location in the media and the incidence coordinates were estimated in the GIS. First, sections of roads and affected towns or municipalities mentioned in the news were selected. Then, a spatial query allowed for the identification of those incidences close to them. Five classes were established, depending on the distance: Class 1, 0–1 km; Class 2, 1–2 km; Class 3, 2–5 km; Class 4, 5–10 km; and Class 5, more than 10 km.
2. The temporal proximity was addressed by analysing the time interval between the date of the news appearing in the local media and the month associated with the previously selected incidences. Five classes were also considered: Class 1, 0–3 months; Class 2, 3–6 months, Class 3, 6–12 months, Class 4, 12–24 months; and Class 5, more than 24 months. Summing the classes for the spatial and temporal proximities, only those incidences with a maximum of 6 points (e.g., Class 3 in both, or Classes 2 and 4 in each one) were selected. Thus, each incidence point could be associated with several rainfall events and their rainfall–duration (E—D) pairs. In addition to the events identified from the news, the complete rainfall series for the two years previous to the month of each incidence were examined, searching for the major events in each interval of duration. If events different from the above were found, they were also added to the database.
3. Finally, the magnitudes of the rainfall events were considered. First, following some previous studies [36,40,51], the E–D pair with the longest return period was selected, for each of the events associated with an incidence, as the one most likely to trigger

it. Moreover, according to [40], of all the possible events (E–D pairs) associated with each incidence, those which presented a return period of fewer than five years were discarded, as they were considered non-relevant for incidence triggering. This procedure allowed for enrichment of the incidence database, thus including several E–D pairs for each incidence (see some examples in the results Section 3.2).

2.5. Rainfall Threshold Calculation

Prior to the determination of the thresholds, the rainfall variables were analysed, by calculating their mean and modal values, both globally and for each of the typologies considered, in order to determine whether there were differences between the landslides of different typologies and magnitudes.

In this work, the determination of thresholds of the rainfall–duration (E–D) type was considered, which usually respond to linear equations of the type:

$$E = a \times D + b, \tag{1}$$

although equations with a power-law can also fit:

$$E = \alpha\, D^{\beta}. \tag{2}$$

This type of threshold has been considered more appropriate for cases where only daily data are available [28,51,52]. They allow for knowledge of the amount of rainfall necessary to generate landslide or erosive processes, depending on the number of days. Nevertheless, thresholds of the intensity–duration (I–D) type were also determined, although these are more commonly used when intensity per hour (mm/hour) data are available [42,43]. In this case, power-law equations were adjusted. Both thresholds were calculated globally for all the incidence points, but they were also discriminated by typologies.

3. Results

3.1. Incidence Database

The incidence database is shown in the map of Figure 3. It shows the typology and magnitude of the incidences, according to published classifications of landslides [77,78], and includes some significant examples in the study area. In general, practically all the incidences corresponded to shallow phenomena but, within them, two types were differentiated, depending on their size or magnitude [79]:

- Very shallow processes, with magnitude between extremely and very small (<5000 m^3). These correspond to ruptures in the road cut, either of the slide or collapse-rockfall typologies, but also undercuts of the road embankment. Meanwhile, erosive processes (gullies) were identified, which also produce incidences on the roads.
- Shallow processes in which there is mobilization of the slope where the road is located, with a magnitude generally between small and medium (5000–500,000 m^3). Within these, slope movements of a slide or flow type were considered, according to [77,78]. Soil creeping processes were also distinguished from those flows which were well-defined in the landscape.

The distribution by typology is shown in Table 2. As can be observed, there were 46 incidences corresponding to gully processes, 47 punctual incidences in road cuts (38 landslides and 9 collapses), and 30 incidences associated to undercutting in road embankments. Then, 77 incidences were directly related to the road; that is, to human activity. Moreover, there were 63 landslides of higher magnitude, among which 21 slides, 26 flows, and 16 creeping areas were identified.

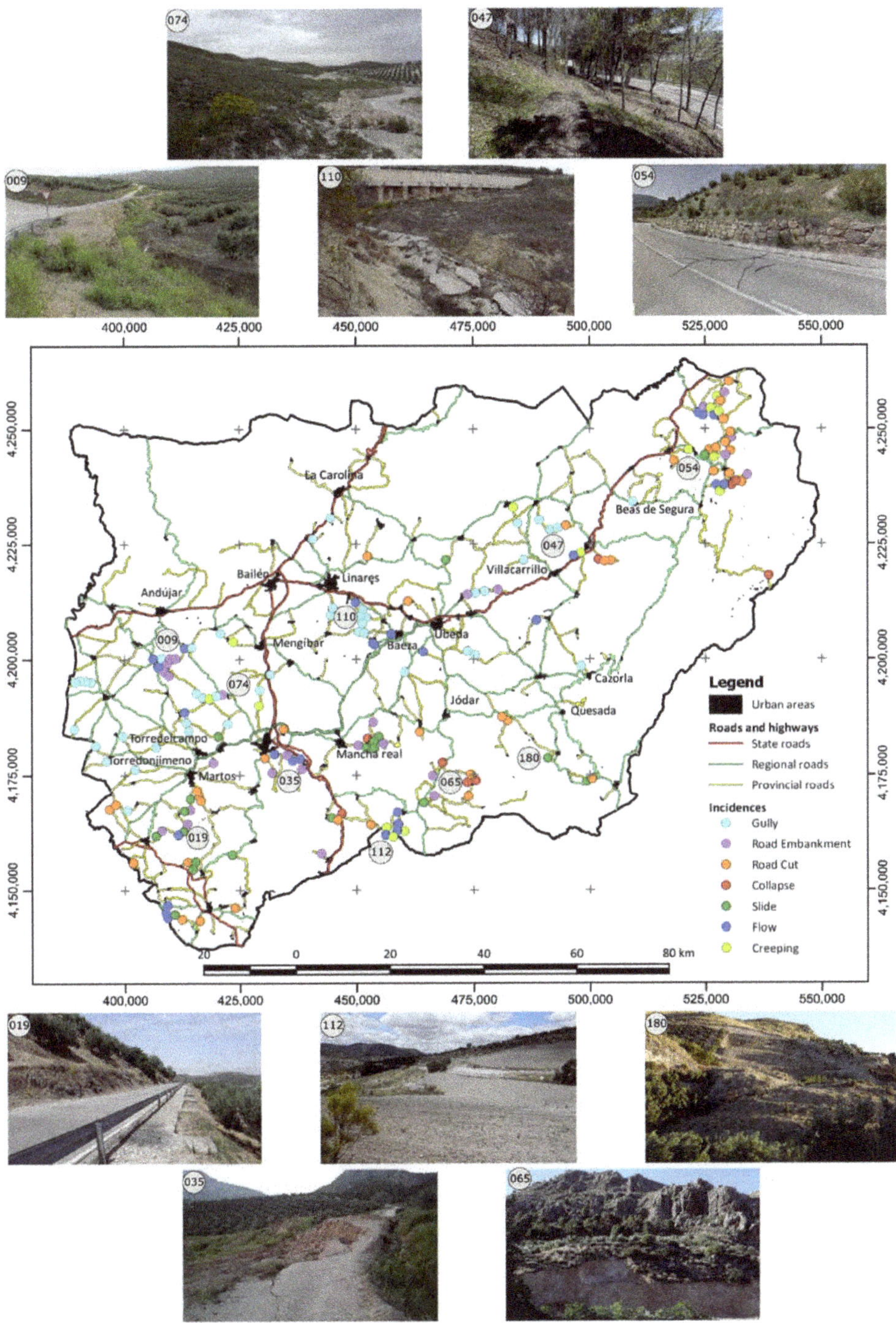

Figure 3. Map of incidences. Points: (009), gully; (019), undercutting of the road embankment; (035), flow; (047), creeping process; (054), slide in the road cut; (065), collapse over the road; (074), creeping process; (110), flow and gullies; (112), flow and creep; (180), slide in the slope of the road.

Table 2. Distribution of incidences by typologies.

Magnitude	Typology	Number
Lower magnitude Very shallow	Gullies	46
	Undercut in road embankments	30
	Slides in road cuts	38
	Collapses in road cuts	9
Higher magnitude Shallow	Slides	21
	Flows	26
	Creep	16

Meanwhile, Table 3 shows the distribution of the year in which the civil works to repair the road started. Most of them were concentrated into two years, 2010 (61 incidences) and 2013 (70 incidences).

Table 3. Distribution of incidences by year in which the civil work started.

Year	Number	Year	Number
1998	2	2006	2
1999	5	2007	1
2000	3	2008	2
2001	3	2009	3
2002	2	2010	61
2003	1	2011	18
2004	3	2012	3
2005	4	2013	70

3.2. *Rainfall Events*

Figure 4 shows the daily rainfall series associated with the aforementioned significant incidence points (shown in the map of Figure 3), where different rainfall events can be observed. Following the methodology described, the rainfall events associated with each incidence were searched in the two years (24 months) previous to the starting of the repair work. Thus, Figure 5 shows the two-year rainfall series for different event durations in two significant incidence points (099 and 181, not shown in Figure 3). Some arrows in red have been included to point out the rainfall events identified in each incidence, which were later used in the thresholds calculation.

A total of 446 rainfall events (E–D pairs) associated with the 186 incidence points were found that met the established criteria of spatial and temporal proximity, as well as the maximum return periods. Thus, 17 points were associated with 4 potential events, 60 points with 3, 86 points with 2, and 21 points with a single event. Meanwhile, some events affected the whole province in a general way and, therefore, the provincial road network, while others affected more restricted sectors. Some of them, those affecting a minimum of 5 points, are shown in Table 4.

Figure 6 also shows isohyets (rainfall) maps of some general events of different dura-tion

Among the general events, those occurring in the autumn–winter of the hydrological year 2009–2010 stood out: 25–30/12/2009, with rainfall close to 150 mm and exceeding 200 mm, in 5 and 15 days, respectively, at about 20–25 points of the provincial network (Figures 5c and 6b); the events of 06–13/01/2010, with rainfall that exceeded 300 mm in 30 days (Figure 6c); and, finally, the accumulated rainfall that occurred on 22/02/2010, when 570 mm was reached in 75 days at 41 points (Figure 6d) and on 02/03/2010 with 660 mm in 90 days (Figure 5i). In all these cases, the return periods were quite long, generally between 15 and 22.5 years and, in some cases, reaching the total period analysed (45 years).

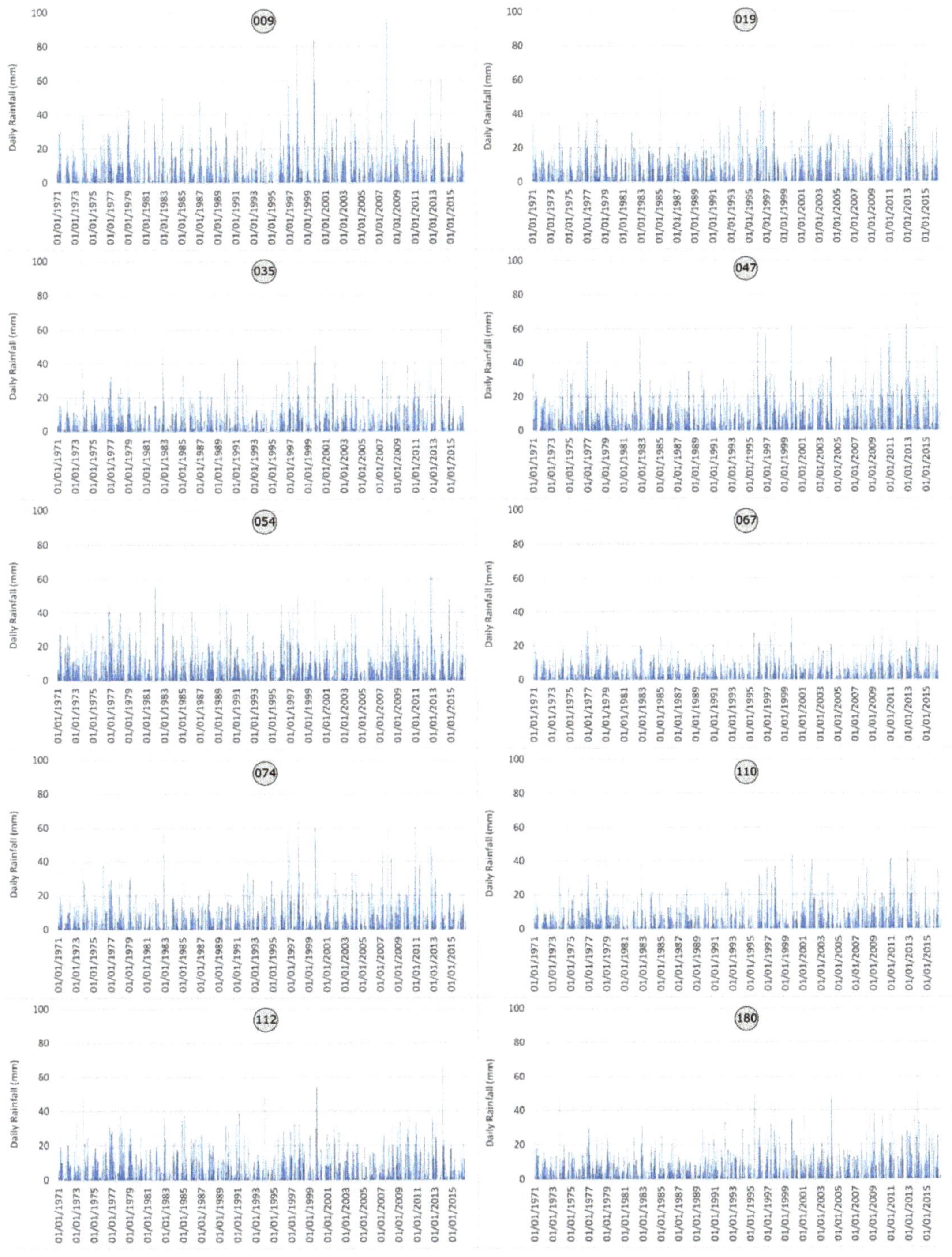

Figure 4. Rainfall series of the incidence points shown in Figure 3.

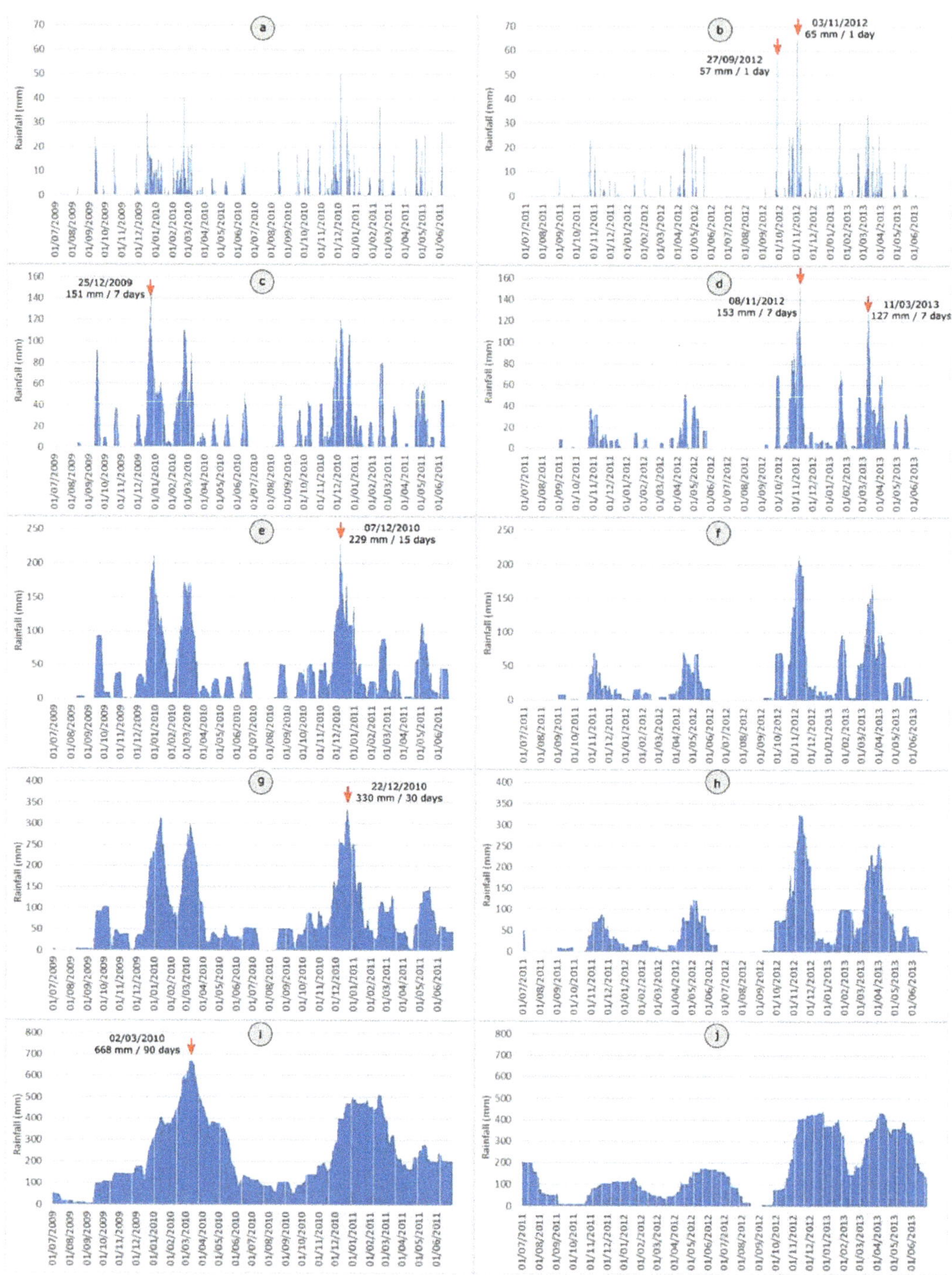

Figure 5. Rainfalls series (two-year) for the incidence points 099 and 174: (**a**,**b**), Daily rainfall; (**c**,**d**), 7-day antecedent rainfall; (**e**,**f**), 15-day antecedent rainfall; (**g**,**h**), 30-day antecedent rainfall; and (**i**,**j**), 90-day antecedent rainfall. Red arrows represent the days (E–D pairs) selected as rainfall events for each point.

Table 4. Main rainfall events associated to incidences. Sector: SS, Sierra Sur; SM, Sierra Mágina; SC, Sierra Cazorla; SG, Sierra Segura; J, Jaén sector; CH, Central Hills; W, Western sector; N, Northern sector; Gen., General.

Date	Number of Points	Mean E (mm)	Modal D (Days)	Mean I (mm/d)	Modal T (Years)	Sector
03/11/1997	5	94.20	2	47.10	15.00	SS, SM, W
31/12/1997	5	411.25	60	6.74	9.00	SS, SC, SG
20/10/1999	6	50.38	1	50.38	22.50	SC, SG, W
28/03/2004	6	75.67	3	27.92	11.25	SS, W, N
08/04/2008	5	50.50	1	50.50	15.00	SS, CH, W
25/12/2009	18	143.29	5	28.66	45.00	Gen.
30/12/2009	26	210.34	15	14.02	22.50	Gen.
06/01/2010	10	268.50	30	8.95	7.50	SS, SM, SC, SG
11/01/2010	7	344.27	30	11.48	22.50	SS, SC, W
13/01/2010	13	324.08	30	10.80	15.00	Gen.
15/02/2010	6	435.90	60	7.15	22.50	SS, SM, CH
22/02/2010	41	568.15	75	7.58	15.00	Gen.
02/03/2010	11	662.56	90	7.36	15.00	Gen.
30/10/2010	6	36.75	1	36.75	5.00	SM, SG, CH
06/12/2010	13	79.40	2	39.70	6.43	SS, SM, SG, N
31/12/2010	5	362.00	45	8.04	11.25	CH
14/02/2011	6	45	1.	37.50	6.43	SS, SG, CH
27/09/2012	15	59.20	1	59.20	11.25	SS, CH, W
03/11/2012	59	72.27	2	36.14	15.00	Gen.
08/11/2012	51	151.38	7	21.63	9.00	Gen.
11/03/2013	43	119.42	7	17.06	5.00	Gen.
18/03/2013	6	166.30	15.	11.09	6.43	SS

The second important rainy period occurred in the year 2012–2013, with several events: 03–08/11/2012, with daily rainfall that exceeded 50 mm, 2-day rainfall around 70 mm, and weekly rainfall that reached 150 mm (Figures 5b–d and 6e,f), all generalized in the road network (50–60 points). The return period was 9 to 15 years. Subsequently, on 18/03/2013, rainfall of 166 mm was reached in 15 days at 43 points, with a return period of 6 years.

More locally, there were other rainfall events potentially associated with incidences such as those at the end of 1997, when 400 mm was exceeded in 60 days (Figure 6a), mainly in the mountain ranges; spring 2004 or 2008, with more than 50 mm in 1 day in different sectors; autumn–winter 2010, with 80 mm in 2 days, 230 mm in 15 days (Figure 5e) and 360 mm in 45 days (Figure 5g) in the mountain ranges and the central hills; or those of the end of summer 2012, with almost 60 mm in a day in the southern and western parts of the province (Figure 5b). In all of these events, the return periods were between 5 and 10 years.

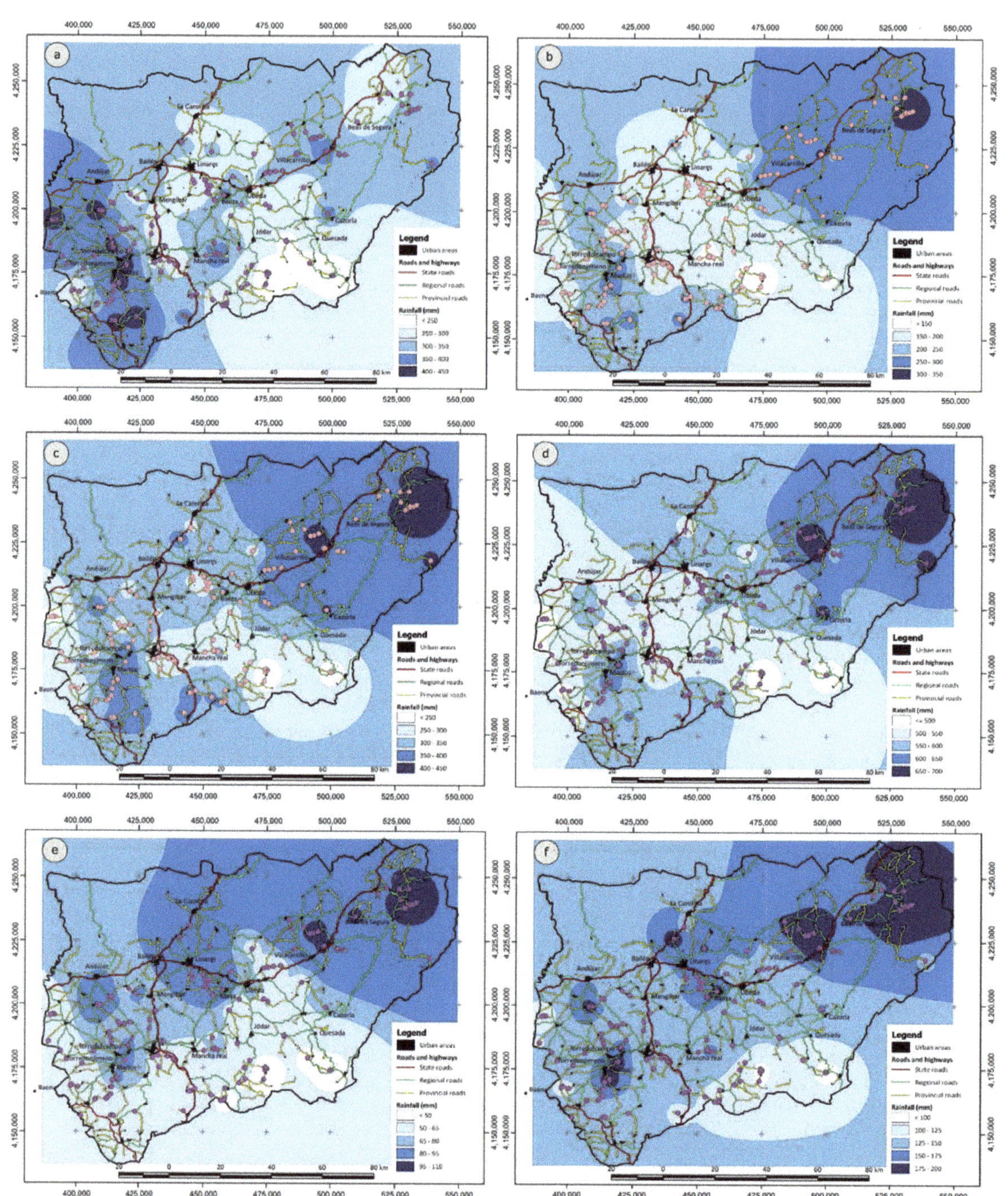

Figure 6. Isohyets maps of some events: (**a**) 31/12/1997 (D = 60 days); (**b**) 30/12/2009 (D = 15 days); (**c**) 13/01/2010 (D = 30 days); (**d**) 22/02/2010 (D = 75 days); (**e**) 03/11/2012 (D = 2 days); and (**f**) 08/11/2012 (D = 7 days).

3.3. Rainfall Thresholds

Table 5 shows the mean values of the rainfall amount, duration, and intensity (E, D and I) variables, as well as the modal value of the duration, the total events, and discriminated by typology. Likewise, Figure 7 shows the histograms of the duration of the events.

Table 5. Mean values of rainfall amount (E), duration (D), and intensity (I) for the rainfall events.

Typology	Mean E (mm)	Mean D (days)	Modal D (days)	Mean I (mm/day)
Gullies	147.33	11.78	1	28.11
Road embankment	126.81	8.84	1	32.55
Slides in road cuts	171.84	15.42	7	26.66
Collapses	209.05	26.45	1	20.49
Lower magnitude (very shallow)	155.29	13.40	1	28.09
Slides	234.79	23.65	1	21.77
Flows	211.68	19.75	7	21.97
Creep	227.87	21.95	7	22.79
Higher magnitude (shallow)	223.40	21.60	7	22.11
Total	178.96	16.25	7	26.01

From the data shown in Table 5 and Figure 7, a shorter duration of rain events was generally observed in the lower magnitude incidences, with an average value of 13.70 days (the modal value being 1 day). Meanwhile, for the higher magnitude incidences, such as slides, flows, and creeping processes, the average duration was 21.60 days with a modal value of 7 days. In the lower magnitude incidences, the average rainfall was 223 mm and the intensity was 22 mm/day. For the higher magnitude landslides, the average rainfall was 155 mm, with an average intensity of 28 mm/day.

Regarding the thresholds, Table 6 shows the equations obtained both for the rainfall–duration (E–D) threshold (linear and power-law adjustment), and for the intensity–duration (I–D) threshold (power-law adjustment). The table also shows the coefficient of determination (R^2) of the adjustment. Figure 8 shows these thresholds for lower and higher magnitude incidences.

As can be seen from Table 6 and Figure 8, the equations were quite similar for both the lower and higher magnitude incidences; although, in the case of linear adjustments, the intercept in the former (72)—especially in the collapses (61)—was lower than that in the latter (82), the slopes being similar. In the same way, the power-law base was somewhat lower in the lower magnitude incidences (47.5) than in the higher magnitude ones (49.7), both for the E–D and I–D thresholds.

Table 6. Equations for E–D (linear and power-law) and I–D thresholds.

Typology	E–D (Linear)		E–D (Power-Law)		I–D (Power-Law)	
	Equation	R^2	Equation	R^2	Equation	R^2
Gullies	$E = 6.294\,D + 73.187$	0.90	$E = 47.283\,D^{0.543}$	0.90	$I = 47.283\,D^{-0.457}$	0.87
Road embankment	$E = 5.985\,D + 73.909$	0.90	$E = 51.155\,D^{0.516}$	0.87	$I = 51.155\,D^{-0.484}$	0.85
Slides in road cuts	$E = 6.586\,D + 70.301$	0.92	$E = 46.313\,D^{0.554}$	0.89	$I = 46.313\,D^{-0.446}$	0.84
Collapses	$E = 5.595\,D + 61.023$	0.85	$E = 39.080\,D^{0.564}$	0.94	$I = 39.080\,D^{-0.436}$	0.90
Very shallow	$E = 6.222\,D + 71.908$	0.90	$E = 47.481\,D^{0.540}$	0.89	$I = 47.481\,D^{-0.460}$	0.86
Slides	$E = 6.793\,D + 74.106$	0.95	$E = 47.089\,D^{0.543}$	0.94	$I = 47.089\,D^{-0.451}$	0.91
Flows	$E = 6.293\,D + 87.366$	0.94	$E = 51.721\,D^{0.523}$	0.93	$I = 51.722\,D^{-0.477}$	0.91
Creeping	$E = 6.488\,D + 85.464$	0.93	$E = 50.537\,D^{0.546}$	0.94	$I = 50.537\,D^{-0.454}$	0.91
Shallow	$E = 6.527\,D + 82.424$	0.94	$E = 49.752\,D^{0.538}$	0.93	$I = 49.752\,D^{-0.462}$	0.91
Total	$E = 6.408\,D + 74.829$	0.92	$E = 47.961\,D^{0.542}$	0.91	$I = 47.961\,D^{-0.458}$	0.88

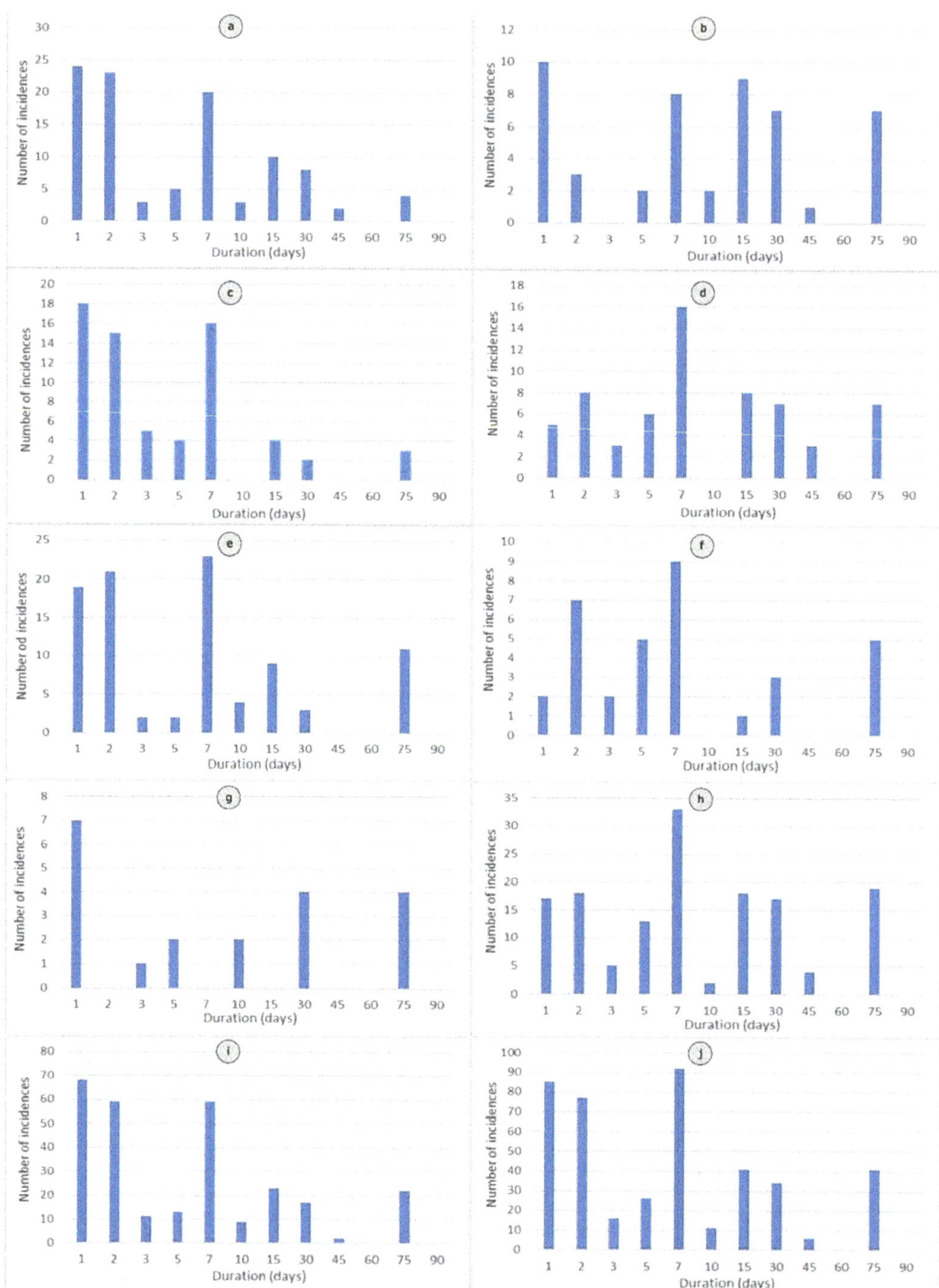

Figure 7. Duration of rainfall event for incidences of the different typologies (**a**): Gullies; (**c**): undercut in road embankments; (**e**): slides in road cuts; (**g**): collapses; (**i**): total of lower magnitude (very shallow) incidences; (**b**): slides; (**d**): flows; (**f**): creep; (**h**): total of higher magnitude (shallow) incidences; and (**j**) all incidences.

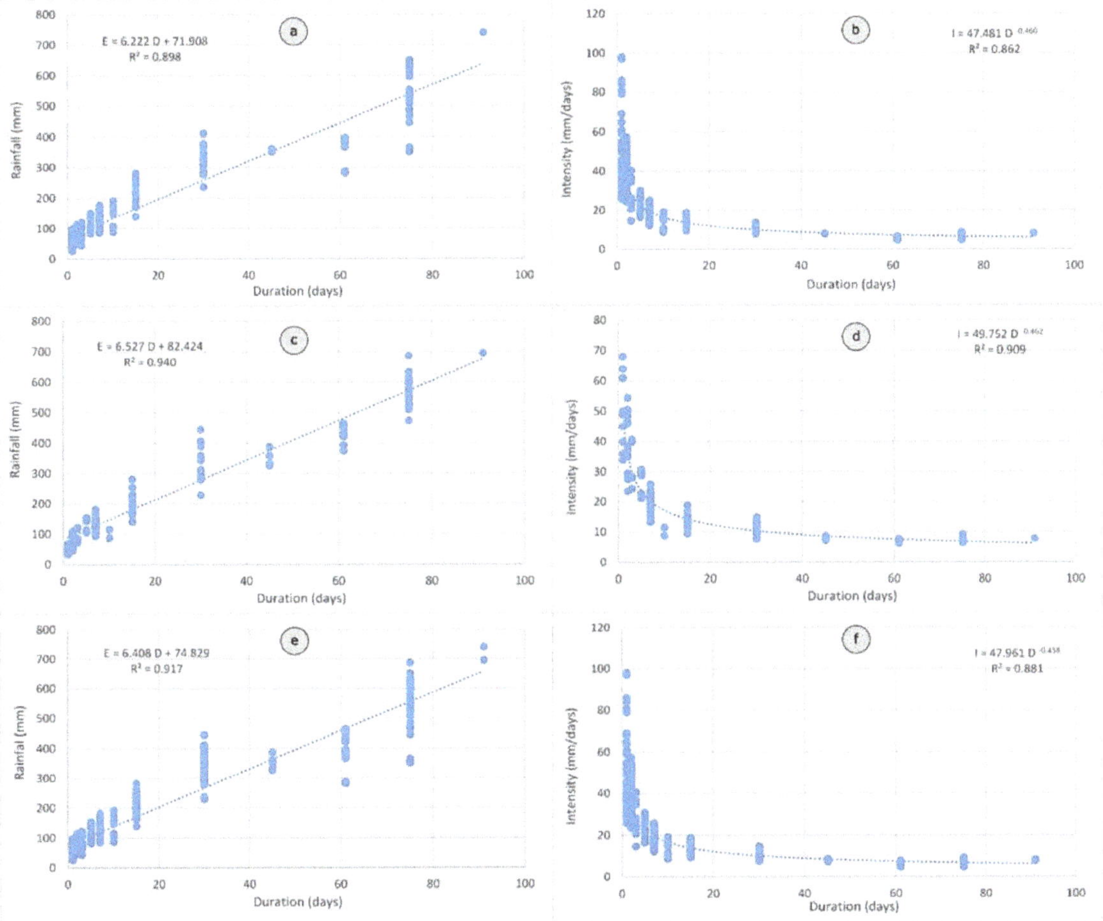

Figure 8. Rainfall thresholds: (**a**) E–D for lower magnitude incidences; (**b**) I–D for lower magnitude incidences; (**c**) E–D for higher magnitude incidences; (**d**) I–D for higher magnitude incidences; (**e**) E–D for all incidences; and (**f**) I–D for all incidences.

4. Discussion

From the results obtained, it can be observed that the road network in the Jaén province was affected by numerous incidences in the period studied. A set of 186 incidences was registered, of which 123 (66%) corresponded to lower magnitude processes (gullies, undercut of road embankments, slides and collapses in road cuts,) and 63 (33%) corresponded to processes that were also shallow but of higher magnitude, affecting the entire slope (slides, flows, and creeping). Although all the movements affected the road network, some of them (undercut on road embankment, slides and collapses in road cuts, comprising 47% of the incidences) can be considered as directly related to human activities.

The temporal distribution was quite irregular, concentrated mainly in two years: 2010 (with 60 incidences) and 2013 (with 71), representing more than 75% of incidences.

This distribution seems to be related to the occurrence of rainy periods in the province. Without the contribution of other factors, given the low tectonic activity [80], rainfall was the main triggering factor for landslides in the province. As indicated in the introduction, the relationship between landslides and rainfall has been well-established throughout the world [32,33,35], particularly in Europe [42,43] and in the Mediterranean countries [8,9,44,46]. In Spain, these relationships have also been found [28,51,52,81].

Thus, the simple observation of the distribution of the mean annual precipitation (MAP) of point 066, representative of the whole province (Figure 1d), as well as the series of daily rainfall associated with different incidence points (Figure 4), allowed for the establishment of this relationship. Thus, considering the mean annual precipitation (MAP) of 533 mm at point 066, 1026 mm was reached in the hydrological year 2009–2010, 911 mm in 2010–2011, and 950 mm in 2012–2013. The activity of these years has also been observed in natural slopes of some sectors of the province [30]. Meanwhile, the remaining years barely exceeded 600 mm, except for 1976–1977, 1995–1998, and 2003–2004. However discarding the first years in which there was no recording of incidences, the intense rainfall of other years, such as 1997–1998 and 2003–2004, was not reflected in the incidence database as explained by the lower magnitude of the rainfall events or because the incidences were not registered (being in the first years of the database elaboration).

Analysis of the rainfall series associated with each incidence made it possible to more precisely identify a set of possible events for each point in different intervals of antecedent rainfall (duration), based on criteria of spatial and temporal proximity, in relation to the news that had appeared in the local media. Figure 5 shows some of these possible events with different durations, associated with two significant incidence points of different magnitude. Thus, for the point 099 (a higher magnitude incidence), several events were observed in the hydrological years 2009–2010 and 2010–2011; while, for the point 174 (a lower magnitude incidence), several events took place in the hydrological year 2012–2013. Some of these events corresponded to those that occurred with different magnitude in some sectors of the province, or those affecting the whole of the road network in a generalized way, as shown in Table 7. They also coincided with the years in which the MAP was higher, as mentioned above.

Table 7. Comparison of the calculated thresholds with those obtained by other authors.

Threshold Type	This Study	Other Studies [1]
E–D linear (mm–days)	$E = 6.228\ D + 69.716$ (low) $E = 6.408\ D + 74.829$ (mean)	$E = 6.21\ D + 90.8$ (low) [51] $E = 6.98\ D + 181.3$ (mean) [51] $E = 4.57\ D + 133$ [81]
E–D power-law (mm–days)	$E = 47.961\ D^{0.542}$	$E = 73.33\ D^{0.76}$ (Ecuador) [52] $E = 52.34\ D^{0.42}$ (Spain) [52]
E–D linear (mm–hours)	$E = 0.267\ D + 74.829$	$E = 70.00 + 0.2625\ D$ [83]
E–D power-law (mm–hours)	$E = 8.557\ D^{0.542}$	$E = 7.7\ D^{0.39}$ [46] $E = 8.6\ D^{0.41}$ [47] $E = 5.6\ D^{0.40}$ [49] $E = 6.0\ D^{0.47}$ [50] $E = 6.1\ D^{0.52}$ [53]
I–D power-law (mm/days–days)	$I = 47.961\ D^{-0.458}$	$I = 88.005\ D^{-0.69}$ [28] $I = 68.645\ D^{-0.593}$ [82] $I = 84.3\ D^{-0.57}$ [40]
I–D power-law (mm/hours–hours)	$I = 8.557\ D^{-0.458}$	$I = 0.48 + 7.2\ D^{-1}$ [15] $I = 9.40\ D^{-0.56}$ [42] $I = 2.20\ D^{-0.44}$ [43] $I = 7.17\ D^{-0.55}$ [44]
IMAP–D (%–hours)	$IMAP = 0.0187\ D^{-0.484}$	$I\ MAP = 0.76\ D^{-0.33}$ [19] $I\ MAP = 0.007\ D^{-0.54}$ [42] $I\ MAP = 0.0016\ D^{-0.40}$ [43]

[1] Some thresholds are average values from the considered studies.

At the point 099 (whose civil works started in June 2011), the two-year antecedent rainfall series began in the hydrological year 2009–2010, with a rainfall of 150 mm in 7 days to more than 200 mm in 15 days in December. Rainfall continued in January, reaching more than 300 mm in 30 days and, in February–March, it exceeded 570 mm in 75 days and 680 mm in 90 days, more than the annual precipitation in most of the province. After summer 2010, the rainfall recovered, reaching values of 120 mm in 7 days, 230 mm in 15 days, and near 340 m in 30 days for December. Thus, four E–D pairs were selected, for different dates that met the aforementioned criteria (Figure 5), with return periods higher than 20 years (even reaching 45 years), as the most important events of the entire rainfall series. Meanwhile, at point 174 (whose civil works started in June 2013), the series of antecedent rainfall reached important values from September and, especially, November 2012, with events of daily rainfall greater than 60 mm and weekly rainfall of 150 mm, accumulating about 300 mm in 30 days by the end of this month. After this rainy period, the rainfall decreased, but recovered in March 2013 and reached values close to 130 mm in 7 days. Then, four E–D pairs were selected, with return periods always greater than 5 years (mostly between 10 and 20 years). These two particular examples coincided with the following analysis, in which the lower magnitude incidences were usually related to intense rainfall of short duration (1–7 days), while higher magnitude incidences required a longer duration (1–3 months).

Analysis of the average values of the considered variables showed some aspects of interest, such as a shorter duration of the events associated with lower magnitude incidences (mode of 1 day and mean of about 13 days), compared to those of higher magnitude (mode of 7 days and mean higher than 20 days). Consequently, the amount of rainfall was lower in the former (around 150 mm) than in the latter (around 225 mm), unlike the intensity (28 and 22 mm/day, respectively). This difference in behaviour has been pointed out in previous studies considering deep landslides [36,40], where the duration of antecedent rainfall ranged between less than 15 days for shallow landslides to more than 30 days for deeper ones [36,40]. Usually, the prediction is more complex for deeper landslides, for which it is necessary to consider the antecedent rainfall that determines the soil moisture conditions in the medium-term [16,18,19,28,40–43,51,52,57,59–62,81], or even the variation in annual rainfall over several years [58]. These landslides are triggered by a reduction in the shear strength of affected soils and rocks, related to the constant increase in groundwater level as an effect of long-term rainfall periods [40]. Although deep landslides were not considered in the strict sense in this study, a certain difference was observed between the lower and higher magnitude incidences, with respect to the duration of events causing the incidences.

In any case, the consideration of antecedent rainfall provides a simple way to introduce hydrogeological conditions into these studies, even when only shallow landslides are analysed. Some studies, which have mainly used daily data (such as in this one), have been based on the analysis of rainfall duration periods longer than 1 month [51], while other, more sophisticated methods distinguish between the antecedent and triggering rainfall [28,52]. The use of a calibrated antecedent rainfall that decreases with time [38,40] provides another way to simulate the hydrological conditions; however, in this preliminary study, in which the incidence date had high uncertainty, a simple method based only on rainfall amount–duration pairs was applied.

Regarding the rainfall thresholds, we focused especially on E–D type pairs, as they are two truly independent variables [45,46,56]. In addition, in this case (as in other ones), we utilised daily data [28,40,51,52,62,82], which can make it difficult to accurately calculate the intensity, with respect to those other studies that use hourly data [42,43]. Nevertheless, the adjustment of linear equations was tested, with good results, as in previous studies with daily data [51], as well as power-law curves for both the E–D and I–D thresholds (Table 6 and Figure 8). The different thresholds show a good general fit (R^2 higher than 0.9), without significant differences between them. Moreover, normalized thresholds using the MAP were calculated, and all the equations were expressed using the duration (D) in hours in order to facilitate comparison with other studies. Most of these results were not included in the previous section, for simplicity, but are summarized in Table 7, where a comparison with the thresholds obtained by other authors is shown.

In general, the thresholds were within an order of magnitude of those corresponding to other studies, validating the results of this work. However, it must be taken into account that some of them used lower value thresholds; that is, thresholds adjusted such that only a reduced percentile (10, 5, or 1%) of the landslides is under the threshold curve. In this study, we mainly considered thresholds adjusted to the mean values, but the threshold of lower values (5%) was also calculated for the linear adjustment (Table 7, low). From both thresholds, the rainfall amount and intensity corresponding to the different durations were calculated, which could be used to develop a warning system regarding the activation of incidences in the road network of the province of Jaén (Table 8).

Table 8. Rainfall and intensity for the different durations considered in the thresholds of mean and lower values.

	1 d	2 d	3 d	5 d	7 d	10 d	15 d	30 d	45 d	60 d	75 d	90 d
E med	81	88	94	107	120	139	171	267	363	459	555	652
E min	76	82	88	101	113	132	163	257	350	443	537	630
I med	81	44	31	21	17	14	11	9	8	8	7	7
I min	76	41	29	20	16	13	11	9	8	7	7	7

Hydro-meteorological thresholds [59–62] are increasingly being used to overcome the drawbacks of rainfall thresholds, which do not consider adequately the hydrological conditions of slopes and, besides, produce a great proportion of false positives in the prediction, thus limiting the development of warning systems. There exist different methods to determine these thresholds, some of them based on hydro(geo)logical models which take into account detailed data of the phreatic level, soil humidity, porosity, permeability, saturation, rainfall, and so on [59–61], while others are based on data at the basin level, such as rainfall, evapotranspiration, runoff, and so on [62]. In both, additional data are necessary, which were not available in this case, such that they were discarded herein.

Considering the differences by typology, only small differences were observed between the values obtained; the intercept of linear adjustment and the base of the power-law being somewhat higher in the incidences of higher magnitude than in the lower magnitude ones. It must be taken into account that, in any case, the incidences affecting the road network were always shallow, and moreover, the temporal resolution of the data (daily rainfall) was most likely too low to find differences. Differences between the incidences more directly related to the human activities regarding to those less related were not observed. The only typology that presented a certain difference with respect to the remaining incidences, were the collapses (with a lower slope and intercept). This typology is usually associated with steeper areas and road cuts (which are prone to landslides) and the thresholds are likely lower than in other incidences. However, as discussed above, the duration showed higher values in the higher magnitude incidences then in the lower magnitude ones.

The importance of assessing the uncertainty in the determination of rainfall thresh-olds has been discussed in several studies, given that they are based on empirical data which are not always acquired with the required accuracy [42–44], both in the spatiotemporal component and in the measurement component (rainfall gauges). Approaches based on bootstrapping techniques [45] have been proposed, in order to estimate the uncertainty of the power-law parameters (α as the base or scaling parameter; and β as the exponent or the shape parameter) in the E–D thresholds. These approaches have been used in other studies oriented to the development of warning systems [45,46,48,49,53], where uncertainty values representing 5–10% of these parameters have been found [50]. The influence of the temporal resolution on the uncertainty has also been analysed, resulting in smaller scaling parameters (intercepts in logarithmic scale), higher shape parameter (slopes), shorter ranges of validity of the thresholds, and higher uncertainties when the temporal resolution decreased [53]. In this study, the main uncertainty source was not the spatial component, as the meteorological data grid was quite dense (1 km) but, instead, the temporal component (daily rainfall resolution), and especially the uncertainty of the incidence date. This date was estimated from the month when the civil work started, applying the criteria of spatial and temporal proximity to the news appearing in the local media, as well as the magnitude of the event (return period). In these conditions, a validation analysis with the use of positive and negative events is difficult to implement. These limitations, in terms of the data, and the preliminary nature of the study, led us to not consider the uncertainty analysis. Nevertheless, future works utilising more data and with the objective of developing a warning system should address these concerns regarding validation and uncertainty analyses.

Finally, a brief note on the relationships between landslide activity, rainfall, and the global climate could be necessary, although it exceeds the objectives of this study. Several studies [36–38,40,52] have pointed to a relationship between rainfall of high intensity (which generates landslides) and global climatic phenomena. Among these, the well-known teleconnections stand out, such as the South Pacific-El Niño oscillation (ENSO), the North Atlantic oscillation (NAO), or the Western Mediterranean oscillation (WeMo). All of them are related, and may have an influence on the rainfall of the Iberian Peninsula and the Mediterranean [84,85], as has been observed in the Balearic Islands [37].

It is well-known that the high negative anomalies of the NAO (NAOi) are the origin of rainy winters and storms in Portugal and the whole of the Iberian Peninsula [84], thus producing floods [86] and landslides [28,36,40,52]. In this case, they are mainly related to deep landslides [40], but also to shallow landslides [36,52]. Meanwhile, the WeMo also seems to influence the rainfall in the Mediterranean area of the Iberian Peninsula [87]. Thus, in southern Spain, the relationships between both indices and rainfall have been analysed [52]. Although a significant correlation has not been obtained, two of the wettest hydrological years, such as 1995–1996 and 2009–2010 (the latter being one of the years with the highest activity in the study area), were observed to be related to negative anomalies of both indices.

Regarding the role of climate change in rainfall-triggered landslides, this is an interesting issue currently in discussion [88], although the prediction of rainfall events that generate landslides has shown a high level of uncertainty [89]. However, it seems clear that a global climate change scenario, in which severe events such as intense rainfalls are expected [90], should influence landslide activity [91], specifically the rainfall thresholds [88]. Predictions for Spain and the Mediterranean [92,93] have pointed out, in this direction, that Severe Weather Threats (SWEATs) are estimated to increase in the near future, especially in the summer and autumn [94]. Thus, higher levels of hazard and risk affecting civil infrastructure in the province of Jaén are also expected, as has been shown by a preliminary approach, in which a 30% increase in daily rainfall was found in some sectors of the province [95].

5. Conclusions

Rainfall thresholds are one of the most widely applied methods for indirectly estimating landslide return periods, which are subsequently used in hazard analyses. They are also used in the implementation of early-warning systems.

In this work, the starting point was an incidence database in the road network of the province of Jaén, in which the positions and dates of civil repair works are contained. Moreover, a meteorological database with daily rainfall data in a dense grid (1 km) allowed us to link the rainfall series to the incidence points accurately. Then, the identification of rainfall events that potentially generated landslides and erosion processes was addressed, using criteria related to the spatial proximity, temporal proximity, and return period, additionally considering the news appearing in the local media.

Regarding the results, some relevant aspects could be extracted:

- Several events were identified, the most important being related to the hydrological years 2009–2010 and 2012–2013. Some of them were located in specific areas and other ones affected practically the entire road network. The return periods of these significant events were always greater than 5 years and, in some cases, exceeded 10–20 years.
- The lower magnitude incidences usually presented a shorter duration (mode of 1–15 days), compared to those of higher magnitude (7–30 days). Consequently, the amount of rain was lower in the former (around 150 mm) than in the latter (around 225 mm).
- The thresholds obtained for both the rainfall–duration (E–D) and intensity–duration (I–D) pairs were on the same order of magnitude as those calculated by other authors some of them in a similar environment (i.e., Mediterranean countries). The different types of thresholds tested (E–D or I–D, linear or power-law) showed a good fit, without significant differences, likely due to duration data being in units of days, not in hours and the shallow nature of all the incidences.
- In this case, there were no differences in the thresholds between the lower and higher magnitude incidences, unlike the variables (E, I, D) themselves.
- Finally, from the thresholds, rainfall amounts and intensities for different durations of the events were calculated (e.g., about 80 mm for 1 day and more than 250 mm for 1 month), considering not only the threshold adjusted to the mean values but the threshold adjusted to the lower values.

Future improvements to the study should first address extension of the database to other roads, and even the inclusion of landslides on natural slopes, which would allow the thresholds and calculated variables to be refined. In this sense, recording the incidences or landslides with precise knowledge of the date of occurrence could contribute to this refinement. The use of more accurate data would also allow for addressing advanced models, in which antecedent and triggering rainfalls are distinguished and, even (if additional data were available), the determination of hydro-meteorological thresholds. Another future approach consists of estimating the uncertainty of the models and their validation (e.g., using different samples of incidences for training and testing, or by means of temporal validation). Then, these thresholds may be reliably used in hazard analysis or for the implementation of (early) warning systems in the study area.

Author Contributions: Conceptualization: R.C., T.F., J.T.-P., M.S.-G. and J.C.; methodology: R.C. T.F., and J.T.-P.; field work: R.C., I.M. and F.M.; data curation: R.C., T.F., I.M. and F.M.; writing: R.C. T.F. and M.S.-G.; supervision: J.T.-P. and M.S.-G.; project administration: R.C. and J.T.-P.; funding acquisition: R.C., J.T.-P. and T.F. All authors have read and agreed to the published version of the manuscript.

Funding: This work was financed by the following projects: The Convene "Risks associated to the Road Network of the Jaén Province" of the Deputy of Jaén; "Development of a methodology for the landslide hazard mapping: Application to the province of Jaén" (Center for Advanced Studies in Earth Sciences, Energy and Environment of the University of Jaén); "SPS-LIDAR (National Research Agency of Spain; ref. RTI2018-099638-B-I00)".

Institutional Review Board Statement: Not applicable.

Informed Consent Statement: Not applicable.

Acknowledgments: The authors thank AEMET and UC for the data provided for this work (Spain02 v5 data set, available at http://www.meteo.unican.es/datasets/spain02 (accessed on 15 May 2021)).We also acknowledge the help of the groups TEP213 (SFT), RNM325, RNM127, and TEP220 (Matras) of PAIDI, and the research project SUSTAINOLIVE (PRIMA projects) of the UE.

Conflicts of Interest: The authors declare no conflict of interest.

References

1. Varnes, D.J. *Landslide Hazard Zonation: A Review of Principles and Practice, Natural Hazards*; UNESCO: Paris, France, 1984.
2. Schuster, R.L. Socioeconomic significance of landslides. In *Landslides: Investigation and Mitigation*; Turner, A.K., Schuster, R.L., Eds.; Transportation Research Board Special Report 247; National Academy of Sciences: Washington, DC, USA, 1996; pp. 12–35.
3. Guzzetti, F.; Carrara, A.; Cardinali, M.; Reichenbach, P. Landslide hazard evaluation: A review of current techniques and their application in a multi-scale study. *Geomorphology* **1999**, *31*, 181–216. [CrossRef]
4. Petley, D. Global patterns of loss of life from landslides. *Geology* **2012**, *40*, 927–930. [CrossRef]
5. GAR. 2009. Available online: https://www.preventionweb.net/english/hyogo/gar/2009/ (accessed on 15 May 2021).
6. GAR. 2019. Available online: https://gar.undrr.org/report-2019 (accessed on 15 May 2021).
7. Spizzichino, D.; Margottini, C.; Trigila, A.; Iadanza, C. Landslide impacts in Europe: Weaknesses and strengths of databases available at European and national scale. In *Landslide Science and Practice, Proceedings of the Second World Landslide Forum, Rome, Italy, 3–9 October 2011*; Springer: Berlin/Heidelberg, Germany, 2013; pp. 73–80.
8. Cardinali, M.; Reichenbach, P.; Guzzetti, F.; Ardizzone, F.; Antonini, G.; Galli, M.; Cacciano, M.; Castellani, M.; Salvati, P. A geomorphological approach to the estimation of landslide hazards and risks in Umbria, Central Italy. *Nat. Hazards Earth Syst. Sci.* **2002**, *2*, 57–72. [CrossRef]
9. Guzzetti, F.; Cardinali, M.; Reichenbach, P.; Cipolla, F.; Sebastiani, C.; Galli, M.; Salvati, P. Landslides triggered by the 23 November 2000 rainfall event in the Imperia Province, Western Liguria, Italy. *Eng. Geol.* **2004**, *73*, 229–245. [CrossRef]
10. Ayala, F.J.; Elizaga, E.; de González Vallejo, L.I. *Impacto Económico y Social de los Riesgos Geológicos en España*; ITGE: Madrid, Spain, 1987; p. 134.
11. Chacón, J.; Irigaray, C.; Fernández, T.; El Hamdouni, R. Engineering geology maps: Landslides and Geographical Information Systems (GIS). *Bull. Eng. Geol. Environ.* **2006**, *65*, 341–411. [CrossRef]
12. Irigaray, C.; Fernández, T.; El Hamdouni, R.; Chacón, J. Evaluation and validation of landslide susceptibility maps obtained by a GIS matrix method: Examples from the Betic Cordillera (southern Spain). *Nat. Hazards* **2007**, *41*, 61–79. [CrossRef]
13. Van Westen, C.J. The modelling of landslide hazards using GIS. *Surv. Geophys.* **2000**, *21*, 241–255. [CrossRef]
14. Terlien, M.T.J. The determination of statistical and deterministic hydrological landslide-triggering thresholds. *Env. Geol.* **1998**, *35*, 124–130. [CrossRef]
15. Crosta, G.B.; Frattini, P. Distributed modelling of shallow landslides triggered by intense rainfall. *Nat. Hazards Earth Syst. Sci.* **2003**, *3*, 81–93. [CrossRef]
16. Bogaard, T.A.; Greco, R. Landslide hydrology: From hydrology to pore pressure. *Wiley Interdiscip. Rev. Water* **2016**, *3*, 439–459. [CrossRef]
17. Lazzari, M.; Piccarreta, M. Landslide disasters triggered by extreme rainfall events: The Case of Montescaglioso (Basilicata, Southern Italy). *Geosciences* **2018**, *8*, 377. [CrossRef]
18. Sidle, R.C.; Greco, R.; Bogaard, T. Overview of landslide hydrology. *Water* **2019**, *11*, 148. [CrossRef]
19. Aleotti, P. A warning system for rainfall-induced shallow failures. *Eng. Geol.* **2004**, *73*, 247–265. [CrossRef]
20. Chung, C.J.F.; Fabbri, A.G. Probabilistic prediction models for landslide hazard mapping. *Photogramm. Eng. Remote Sens.* **1999**, *65*, 1389–1399.
21. Reichenbach, P.; Rossi, M.; Malamud, B.; Mihri, M.; Guzzetti, F. A review of statistically-based landslide susceptibility models. *Earth Sci. Rev.* **2018**, *180*, 60–91. [CrossRef]
22. Carrara, A. Multivariate models for landslide hazard evaluation. A "Black Box" approach. In Proceedings of the Workshop on Natural Disasters in European Mediterranean Countries, Perugia, Italy, 27 June–1 July 1998; pp. 205–224.
23. Brenning, A. Spatial prediction models for landslide hazards: Review, comparison and evaluation. *Nat. Hazards Earth Syst. Sci.* **2005**, *5*, 853–862. [CrossRef]
24. Merghadi, A.; Yunus, A.P.; Dou, J.; Whiteley, J.; Thaipham, B.; Tien, D.; Avtar, R.; Abderrahmane, B. Earth-science reviews machine learning methods for landslide susceptibility studies: A comparative overview of algorithm performance. *Earth Sci. Rev.* **2020**, *207*, 103225. [CrossRef]
25. Guzzetti, F.; Cesare, A.; Cardinali, M.; Fiorucci, F.; Santangelo, M.; Chang, K. Earth-science reviews landslide inventory maps: New tools for an old problem. *Earth Sci. Rev.* **2012**, *112*, 42–66. [CrossRef]
26. Fell, R.; Corominas, J.; Bonnard, C.; Cascini, L.; Leroi, E.; Savage, W.Z. Guidelines for landslide susceptibility, hazard and risk zoning for land use planning. *Eng. Geol.* **2008**, *102*, 85–98. [CrossRef]
27. Ibsen, M.-L.; Brunsden, D. The nature, use and problems of historical archives for the temporal occurrence of landslides, with specific reference to the south coast of Britain, Ventnor, Isle of Wight. *Geomorphology* **1996**, *15*, 241–258. [CrossRef]
28. Palenzuela, J.A.; Jiménez-Perálvarez, J.D.; Chacón, J.; Irigaray, C. Assessing critical rainfall thresholds for landslide triggering by generating additional information from a reduced database: An approach with examples from the Betic Cordillera (Spain). *Nat. Hazards* **2016**, *84*, 185–212. [CrossRef]
29. Scaioni, M.; Longoni, L.; Melillo, V.; Papini, M. Remote sensing for landslide investigations: An overview of recent achievements and perspectives. *Remote Sens.* **2014**, *6*, 9600–9652. [CrossRef]
30. Fernández, T.; Pérez-García, J.L.; Gómez-López, J.M.; Cardenal, J.; Moya, F.; Delgado, J. Multitemporal landslide inventory and activity analysis by means of aerial photogrammetry and LiDAR techniques in an area of Southern Spain. *Remote Sens.* **2021**, *13*, 2110. [CrossRef]

31. Corominas, J.; Moya, J. A review of assessing landslide frequency for hazard zoning purposes. *Eng. Geol.* **2008**, *102*, 193–213. [CrossRef]
32. Crozier, M.J. Techniques for the morphometric analysis of landslips. *Z. Geomorphol.* **1973**, *17*, 78–101.
33. Finlay, P.J.; Fell, R.; Maguire, P.K. The relationship between the probability of landslide occurrence and rainfall. *Can. Geotech. J.* **1997**, *34*, 811–824. [CrossRef]
34. Irigaray, C.; Lamas, F.; El Hamdouni, R.; Fernández, T.; Chacón, J. The importance of the precipitation and the susceptibility of the slopes for the triggering of landslides along the roads. *Nat. Hazards* **2000**, *21*, 65–81. [CrossRef]
35. Wieczorek, G.; Glade, T. Climatic factors influencing occurrence of debris flows. In *Debris-Flow Hazards and Related Phenomena*; Springer: Berlin/Heidelberg, Germany, 2005; pp. 325–362.
36. Zezere, J.L.; Trigo, R.M.; Fragoso, M.; Oliveira, S.C.; Garcia, R.A.C. Rainfall-triggered landslides in the Lisbon region over 2006 and relationships with the North Atlantic Oscillation. *Nat. Hazards Earth Syst. Sci.* **2008**, *8*, 483–499. [CrossRef]
37. Luque-Espinar, J.A.; Mateos, R.M.; García-Moreno, I.; Pardo-Igúzquiza, E.; Herrera, G. Spectral analysis of climate cycles to predict rainfall induced landslides in the western Mediterranean. *Nat. Hazards* **2017**, *89*, 985–1007. [CrossRef]
38. Moreiras, S.M.; Vergara, I.; Pont, D.; Araneo, D. Were merely storm-landslides driven by the 2015–2016 Niño in the Mendoza River valley? *Landslides* **2018**, *15*, 997–1014. [CrossRef]
39. Caine, N. The rainfall intensity: Duration control of shallow landslides and debris flows. *Geogr. Ann. Ser. A Phys. Geogr.* **1980**, *62*, 23–27.
40. Zezere, J.L.; Trigo, R.; Trigo, I. Shallow and deep landslides induced by rainfall in the Lisbon region (Portugal): Assessment of relationships with the North Atlantic Oscillation. *Nat. Hazards Earth Syst. Sci.* **2005**, *5*, 331–344. [CrossRef]
41. Cardinali, M.; Galli, M.; Guzzetti, F.; Ardizzone, F.; Reichenbach, P.; Bartoccini, P. Rainfall induced landslides in December 2004 in south-western Umbria, central Italy: Types, extent, damage and risk assessment. *Nat. Hazards Earth Syst. Sci.* **2006**, *6*, 237–260. [CrossRef]
42. Guzzetti, F.; Peruccacci, S.; Rossi, M.; Stark, C.P. Rainfall thresholds for the initiation of landslides in central and southern Europe. *Meteorol. Atmos. Phys.* **2007**, *98*, 239–267. [CrossRef]
43. Guzzetti, F.; Peruccacci, S.; Rossi, M.; Stark, C. The rainfall intensity—Duration control of shallow landslides and debris flows: An update. *Landslides* **2008**, *5*, 3–17. [CrossRef]
44. Brunetti, M.T.; Peruccacci, S.; Rossi, M.; Luciani, S.; Valigi, D.; Guzzetti, F. Rainfall thresholds for the possible occurrence of landslides in Italy. *Nat. Hazards Earth Syst. Sci.* **2010**, *10*, 447–458. [CrossRef]
45. Peruccacci, S.; Brunetti, M.T.; Luciani, S.; Vennari, C.; Guzzetti, F. Lithological and seasonal control of rainfall thresholds for the possible initiation of landslides in central Italy. *Geomorphology* **2012**, *139-140*, 79–90. [CrossRef]
46. Peruccacci, S.; Brunetti, M.T.; Gariano, S.L.; Melillo, M.; Rossi, M.; Guzzetti, F. Rainfall thresholds for possible landslide occurrence in Italy. *Geomorphology* **2017**, *290*, 39–57. [CrossRef]
47. Vennari, C.; Gariano, S.L.; Antronico, L.; Brunetti, M.T.; Iovine, G.; Peruccacci, S.; Terranova, O.; Guzzetti, F. Rainfall thresholds for shallow landslide occurrence in Calabria, Southern Italy. *Nat. Hazards Earth Syst. Sci.* **2014**, *14*, 317–330. [CrossRef]
48. Melillo, M.; Brunetti, M.T.; Peruccacci, S.; Gariano, S.L.; Guzzetti, F. Rainfall thresholds for the possible landslide occurrence in Sicily (Southern Italy) based on the automatic reconstruction of rainfall events. *Landslides* **2016**, *13*, 165–172. [CrossRef]
49. Melillo, M.; Brunetti, M.T.; Peruccacci, S.; Gariano, S.L.; Roccati, A.; Guzzetti, F. A tool for the automatic calculation of rainfall thresholds for landslide occurrence. *Environ. Model. Softw.* **2018**, *105*, 230–243. [CrossRef]
50. Palladino, M.R.; Viero, A.; Turconi, L.; Brunetti, M.T.; Peruccacci, S.; Melillo, M.; Luino, F.; Deganutti, A.M.; Guzzetti, F. Rainfall thresholds for the activation of shallow landslides in the Italian Alps: The role of environmental conditioning factors. *Geomorphology* **2018**, *303*, 53–67. [CrossRef]
51. Valenzuela, P.; Zêzere, J.L.; Domínguez-Cuesta, M.J.; Antonio, M.; García, M. Empirical rainfall thresholds for the triggering of landslides in Asturias (NW Spain). *Landslides* **2019**, *16*, 1285–1300. [CrossRef]
52. Palenzuela, J.A.; Soto, J.; Irigaray, C. Characteristics of rainfall events triggering landslides in two climatologically different areas: Southern Ecuador and southern Spain. *Hydrology* **2020**, *7*, 45. [CrossRef]
53. Gariano, S.L.; Melillo, M.; Peruccacci, S.; Brunetti, M.T. How much does the rainfall temporal resolution affect rainfall thresholds for landslide triggering? *Nat. Hazards* **2020**, *100*, 655–670. [CrossRef]
54. IRPI. Available online: http://rainfallthresholds.irpi.cnr.it/threshold_info.htm (accessed on 15 May 2021).
55. Daggupati, P.; Douglas-Mankin, K.; Sheshukov, A.; Barnes, P. Monitoring and estimating ephemeral gully erosion using field measurements and GIS. In Proceedings of the ASABE Annual International Meeting, Pittsburgh, PA, USA, 20–23 June 2010; p. 1009663.
56. Swiechowicz, J. Rainfall threshold for ephemeral gully erosion in foothill cultivated lands (Wisnicz Foothills, Poland). In Proceedings of the EGU General Assembly, Vienna, Austria, 17–22 April 2016; Volume 18.
57. Glade, T.; Crozier, M.; Smith, P. Applying probability determination to refine landslide-triggering rainfall thresholds using an empirical "Antecedent Daily Rainfall Model". *Pure Appl. Geophys.* **2000**, *157*, 1059–1079. [CrossRef]
58. Bonnard, C.H.; Noverraz, F. Influence of climate change on large landslides: Assessment of long term movements and trends. In Proceedings of the International Conference on Landslides—Causes, Impacts and Countermeasures, Davos, Switzerland, 17–21 June 2001; Kuhne, M., Einstein, H.H., Krauter, E., Klapperich, H., Pottler, R., Eds.; VGE: Essen, Germany, 2001; pp. 121–138.

59. Bogaard, T.A.; Greco, R. Invited perspectives: Hydrological perspectives on precipitation intensity-duration thresholds for landslide initiation: Proposing hydro-meteorological thresholds. *Nat. Hazards Earth Syst. Sci.* **2018**, *18*, 31–39. [CrossRef]
60. Marino, P.; Peres, D.J.; Cancelliere, A.; Greco, R.; Bogaard, T.A. Soil moisture information can improve shallow landslide forecasting using the hydrometeorological threshold approach. *Landslides* **2020**, *17*, 2041–2054. [CrossRef]
61. Mirus, B.B.; Morphew, M.D.; Smith, J.B. Developing hydro-meteorological thresholds for shallow landslide initiation and early warning. *Water* **2018**, *10*, 1274. [CrossRef]
62. Ciavolella, M.; Bogaard, T.A.; Gargano, A.; Greco, R. Is there predictive power in hydrological catchment information for regional landslide hazard assessment? *Procedia Earth Planet. Sci.* **2016**, *16*, 195–203. [CrossRef]
63. Martelloni, G.; Segoni, S.; Fanti, R.; Catani, F. Rainfall thresholds for the forecasting of landslide occurrence at regional scale. *Landslides* **2012**, *9*, 485–495. [CrossRef]
64. Segoni, S.; Lagomarsino, D.; Fanti, R.; Moretti, S.; Casagli, N. Integration of rainfall thresholds and susceptibility maps in the Emilia Romagna (Italy) regional-scale landslide warning system. *Landslides* **2015**, *12*, 773–785. [CrossRef]
65. Lin, G.; Chang, M.; Huang, Y.; Ho, J. Assessment of susceptibility to rainfall-induced landslides using improved self-organizing linear output map, support vector machine, and logistic regression. *Eng. Geol.* **2017**, *224*, 62–74. [CrossRef]
66. Pérez-Valera, F.; Sánchez-Gómez, M.; Pérez-López, A.; Pérez-Valera, L.A. An evaporite-bearing accretionary complex in the northern front of the Betic-Rif orogeny. *Tectonics* **2017**, *36*, 1006–1036. [CrossRef]
67. AEMET. Agencia Estatal de Meteorología, 2018. Mapas Climáticos de España (1981–2010). Ministerio para la Transición Ecológica, España. Available online: http://www.aemet.es/documentos/es/conocermas/recursos_en_linea/publicaciones_y_estudios/publicaciones/MapasclimaticosdeEspana19812010/MapasclimaticosdeEspana19812010.pdf (accessed on 15 May 2021).
68. Martín-Vide, J.; Olcina, J. *Climas y Tiempos de España*; Alianza Editorial: Madrid, Spain, 2001; p. 258.
69. ESYRCE. Encuesta Sobre Superficies y Rendimientos de Cultivos. *Ministerio de Agricultura, Pesca y Alimentación. España.* 2019. Available online: https://www.mapa.gob.es/es/estadistica/temas/estadisticas-agrarias/agricultura/esyrce/ (accessed on 15 May 2021).
70. INE. Instituto Nacional de Estadística. Available online: https://www.ine.es/jaxiT3/Tabla.htm?t=2876&L=0 (accessed on 15 May 2021).
71. Herrera, S.; Gutiérrez, J.M.; Ancell, R.; Pons, M.R.; Frías, M.D.; Fernández, J. Development and analysis of a 50 year high-resolution daily gridded precipitation dataset over Spain (Spain02). *Int. J. Climatol.* **2012**, *32*, 74–85. [CrossRef]
72. Herrera, S.; Fernández, J.; Gutiérrez, J.M. Update of the Spain02 gridded observational dataset for Euro-CORDEX evaluation: Assessing the effect of the interpolation methodology. *Int. J. Climatol.* **2016**, *36*, 900–908. [CrossRef]
73. RIA Database. Available online: https://www.juntadeandalucia.es/agriculturaypesca/ifapa/riaweb/web/inicio_estaciones (accessed on 15 May 2021).
74. SAIH. Available online: https://www.chguadalquivir.es/saih/ (accessed on 15 May 2021).
75. MATRAS. Available online: https://matras.ujaen.es/ (accessed on 15 May 2021).
76. IDEAL. Available online: https://www.ideal.es/hemeroteca/historico.html (accessed on 15 May 2021).
77. Varnes, D.J. Slope movement, types and processes. In *Landslides: Analysis and Control*; Schuster, R.L., Krizek, R.J., Eds.; Transportation Research Board Special Report; National Academy of Sciences: Washington, DC, USA, 1978; Volume 176, pp. 12–33.
78. Hungr, O.; Leroueil, S.; Picarelli, L. The Varnes classification of landslide types, an update. *Landslides* **2014**, *11*, 167–194. [CrossRef]
79. Fell, R. Landslide risk assessment and acceptable risk. *Can. Geotech. J.* **1994**, *31*, 261–272. [CrossRef]
80. Sánchez-Gómez, M.; Peláez, J.A.; García-Tortosa, F.J.; Pérez-Valera, F.; Sanz de Galdeano, C. La serie sísmica de Torreperogil (Jaén, Cuenca del Guadalquivir oriental): Evidencias de deformación tectónica en el área epicentral. *Rev. Soc. Geol. Esp.* **2014**, *27*, 301–318.
81. Corominas, J.; Moya, J. Reconstructing recent landslide activity in relation to rainfall in the Llobregat River basin, Eastern Pyrenees, Spain. *Geomorphology* **1999**, *30*, 79–93. [CrossRef]
82. Floris, M.; Mari, M.; Romeo, R.W.; Gori, U. Modelling of landslide-triggering factors—A case study in the Northern Apennines, Italy. In *Lecture Notes in Earth Sciences: Engineering Geology for Infrastructure Planning in Europe*; Hack, R., Azzam, R., Charlier, R., Eds.; Springer: Berlin/Heidelberg, Germany, 2004; Volume 104, pp. 745–753.
83. Zezere, J.L.; Rodrigues, M.L. Rainfall thresholds for landsliding in Lisbon Area (Portugal). In *Landslides*; Rybar, J., Stemberk, J., Wagner, P., Eds.; A.A. Balkema: Lisse, The Netherlands, 2002; pp. 333–338.
84. Trigo, R.M.; Pozo, D.; Timothy, C.; Osborn, J.; Castro, Y.; Gámiz, S.; Esteban, M.J. NAO influence on precipitation, river flow and water resources in the Iberian Peninsula. *Int. J. Climatol.* **2004**, *24*, 925–944. [CrossRef]
85. Pozo-Vázquez, D.; Gámiz-Fortis, S.R.; Tovar-Pescador, J.; Esteban-Parra, M.J.; Castro-Díez, Y. El Niño—Southern Oscillation events and associated European winter precipitation anomalies. *Int. J. Climatol.* **2005**, *25*, 17–31. [CrossRef]
86. Santos, M.; Fragoso, M.; Santos, J.A. Damaging flood severity assessment in Northern Portugal over more than 150 years (1865–2016). *Nat. Hazards* **2018**, *91*, 983–1002. [CrossRef]
87. Lemus Cánovas, M.; López-Bustins, J.A. Variabilidad espacio-temporal de la precipitación en el sur de Cataluña y su relación con la oscilación del Mediterráneo Occidental (WeMO). In Proceedings of the X International Congress AEC: Clima, Sociedad, Riesgos y Ordenación del Territorio, Alicante, Spain, 5–8 October 2016; Volume 21, pp. 225–236.
88. Alvioli, M.; Melillo, M.; Guzzetti, F.; Rossi, M.; Palazzi, E.; von Hardenberg, J.; Brunetti, M.T.; Peruccacci, S. Implications of climate change on landslide hazard in Central Italy. *Sci. Total Environ.* **2018**, *630*, 1528–1543. [CrossRef] [PubMed]

89. Coe, J.A.; Godt, J.W. Review of approaches for assessing the impact of climate change on landslide hazards. In Proceedings of the Landslides and Engineered Slopes, Protecting Society through Improved Understanding 11th International and 2nd North American Symposium on Landslides and Engineered Slopes, Banff, AB, Canada, 3–8 June 2012; pp. 371–377.
90. IPCC. *Climate Change 2013: The Physical Science Basis. Contribution of Working Group I to the Fifth Assessment Report of the Intergovernmental Panel on Climate Change*; Cambridge University Press: Cambridge, UK; New York, NY, USA, 2013.
91. Gariano, S.L.; Guzzetti, F. Landslides in a changing climate. *Earth Sci. Rev.* **2016**, *162*, 227–252. [CrossRef]
92. Lin-Ye, J.; García-León, M.; Gracia, V.; Ortego, M.I.; Conte, D.; Perez-Gomez, B.; Sánchez-Arcilla, A. Modelling of future extreme storm surges at the NW Mediterranean coast (Spain). *Water* **2020**, *12*, 472. [CrossRef]
93. Mathbout, S.; Lopez-Bustins, J.A.; Royé, D.; Martin-Vide, J.; Benhamrouche, A. Spatiotemporal variability of daily precipitation concentratio and its relationship to teleconnection patterns over the Mediterranean during 1975–2015. *Int. J. Climatol.* **2019**, *40*, 1435–1455. [CrossRef]
94. Viceto, C.; Marta-Almeida, M.; Rocha, A. Future climate change of stability indices for the Iberian Peninsula. *Int. J. Climatol.* **2017**, *37*, 4390–4408. [CrossRef]
95. Fernández, T.; Gómez-López, J.M.; Pérez-García, J.L.; Cardenal, J.; Delgado, J.; Tovar-Pescador, J.; Sánchez-Gómez, M.; Calero, J. Analysis of the evolution of gully erosion in olive groves using photogrammetry techniques. Relationships with rainfall regime. *ISPRS Ann. Photogramm. Remote Sens. Spat. Inf. Sci.* **2020**, *VI-3/W1-2020*, 19–26. [CrossRef]

Article

Statistical Analysis of Landslide Susceptibility, Macerata Province (Central Italy)

Matteo Gentilucci *, Marco Materazzi and Gilberto Pambianchi

Geology Division, School of Science and Technology, University of Camerino, 62032 Camerino, Italy; marco.materazzi@unicam.it (M.M.); gilberto.pambianchi@unicam.it (G.P.)
* Correspondence: matteo.gentilucci@unicam.it

Abstract: Every year, institutions spend a large amount of resources to solve emergencies generated by hydrogeological instability. The identification of areas potentially subject to hydrogeological risks could allow for more effective prevention. Therefore, the main aim of this research was to assess the susceptibility of territories where no instability phenomena have ever been detected. In order to obtain this type of result, statistical assessments of the problem cannot be ignored. In this case, it was chosen to analyse the susceptibility to landslide using a flexible method that is attracting great interest in the international scientific community, namely the Weight of Evidence (WoE). This model-building procedure, for calculating landslide susceptibility, used Geographic Information Systems (GIS) software by means of mathematical operations between rasters and took into account parameters such as geology, acclivity, land use, average annual precipitation and extreme precipitation events. Thus, this innovative research links landslide susceptibility with triggering factors such as extreme precipitation. The resulting map showed a low weight of precipitation in identifying the areas most susceptible to landslides, although all the parameters included contributed to a more accurate estimate, which is necessary to preserve human life, buildings, heritage and any productive activity.

Keywords: GIS; weight of evidence; susceptibility map; landslides; extreme precipitation

Citation: Gentilucci, M.; Materazzi, M.; Pambianchi, G. Statistical Analysis of Landslide Susceptibility, Macerata Province (Central Italy). *Hydrology* **2021**, *8*, 5. https://doi.org/10.3390/hydrology8010005

Received: 1 December 2020
Accepted: 5 January 2021
Published: 7 January 2021

Publisher's Note: MDPI stays neutral with regard to jurisdictional claims in published maps and institutional affiliations.

1. Introduction

1.1. State of the Art

The Italian territory is subject to a high level of hydrogeological instability and also the province of Macerata is no exception with 7.3% [1] of the territory affected by landslide hazard of grade 3 and 4, where 4 represents the maximum hazard. It follows that landslides susceptibility, which is the statistical likelihood of a landslide occurring in an area, is a very important issue that needs to be studied in depth, also because of the huge resources that are absorbed to deal with emergencies. In this context, climate change is exacerbating the hydrogeological risk and this influence has been demonstrated in numerous studies [2,3]. Hydrogeological risk determines the risk related to the instability of slopes, due to particular geological and geomorphological aspects of these, or of watercourses due to the particular environmental conditions, with possible consequences on the safety of the population and the safety of services and activities on a given territory. Climate change is a trigger for increased hydrogeological risk, it is largely generated by an increase in greenhouse gases which absorbs heat and retain it by gradually releasing it [4], this energy growth, affects both precipitation and temperature. Obviously, it would be useful to work in upstream using countermeasures to contrast climate change by reducing CO_2 emissions or reusing them [5]. However, it is necessary to take note of the current situation, where climate is increasingly the crucial issue. Recently, a lot of research has been carried out to study the impact of climate change on hydrogeological risk, especially landslides [6], although there are other factors that greatly influence terrain stability, such as land use [7]. It is precisely land use that can cause an amplification of the possibility of landslides due to the increase in erosion caused by anthropogenic changes, but also to natural phenomena such as the growth of vegetation or the properties of the soil itself [8]. Other factors influencing landslides include slope [9], lithology [10] and seismic risk [11]; these parameters which contribute to hydrogeological instability lead us to introduce another concept, that of "susceptibility". Landslide susceptibility is the probability that a landslide will occur in a territory, depending on local conditions. It is a measure of the degree to which a territory may be affected by landslides, i.e., an estimate of "where" landslides may occur. There have been many attempts to obtain a probabilistic statistical model, that can allow a reliable assessment of susceptibility [12,13]. The comparison of statistical models for susceptibility calculation (certainty factor, weight of evidence, analytic hierarchy process, etc.) is aimed at defining the best model that allows a minimization of errors, based on landslides detected but deliberately not included in the model-building procedure [14,15]. In this case, the excellent results achieved in scientific literature by the Weight of Evidence (WoE), led to consider it as a reference model for this study. The WoE was originally introduced to assist mining research in identifying new deposits or more accurate reserve estimates [16,17]. The application of this method with the help of Geographic Information Systems (GIS) in the same way, has always been due to applications related to mining research [16]. The maps produced by GIS with the WoE methods, allow areas to be discriminated on the basis of factors that produce certain eventualities. In recent years, this method has been widely applied to landslides as a forecasting tool, with the help of GIS software all over the world [18,19]. Similarly, in Italy, this method has been considered and tests have been carried out in very localized areas [20] and in the mountainous areas of the Apennines and the Alps [21,22]. However the major problem in creating landslide susceptibility maps, is represented by a complete sampling of the factors that can cause instability. Most of the studies are based on small portions of homogeneous territory that obviously cannot be representative of the total and above all, that show many different combinations for example of lithologies, soils, land uses, etc. Instead, this study aims to sample a very large area, carrying out an analysis of the whole territory of the Macerata province, in central Italy. In this area, no studies have been carried out, using WoE and GIS software to obtain a susceptibility map. In any case, the most innovative part of this research lies in the inclusion of the extreme precipitation events, among the parameters that can cause instability. Therefore, this research could represent a link between a study on landslide susceptibility and a study on trigger thresholds. In fact, one of the factors triggering landslides is frequently rainfall, so it is essential to carry out in-depth climatic analyses of the area under investigation [23–25]. An in-depth analysis was carried out, in terms of variation and magnitude of average and extreme rainfall. [26]. Increasingly frequent extreme events dictated by climate change [27] lead to continuous adjustments of susceptibility maps. Forecasting areas of potential instability is of great interest firstly for the protection of human life, and secondly for the cost associated with emergency management. Furthermore, in this area of Italy there are valuable crops, such as vines [28], which can be adversely affected by slope instability and which must be protected to avoid economic consequences.

1.2. Study Area

The study area is the province of Macerata, it is located in central Italy and overlooking the Adriatic Sea, which is part of the Mediterranean Sea. The area is about 2779 Km^2, 67% of the territory is hilly and the remaining 33% is mountainous. To the west, the territory of the province of Macerata (Figure 1) is bordered by the Sibillini mountains (South-western side of the province), part of the Apennine chain, which reach peaks higher than 2200 m a.s.l. Going eastwards there is a wide range of hills that gradually slopes down to the Adriatic coast. Almost all the rivers in the area have a west-east direction except for the Nera river, which crosses the municipalities of Visso and Castelsantangelo sul Nera, and one of its tributaries, the Ussita, both flowing into the Tyrrhenian Sea after joining the Tevere river (Figure 1). From a morphogenetic point of view, the structure of the Umbria-Marche

Appenines is dominated by thrust faults, due to the collisional movement of the African tectonic plate with the European one, while in some internal areas (Tuscany) in the same period (Middle Miocene) there was an extensional tectonic and both are still active. The Umbria-Marche Appenines show an arc with East-facing convexity where it is possible to observe internal wrinkle ridges, an intermediate complex of synclines and external wrinkle ridges.The internal wrinkle ridges consist of various asymmetrical east-vergent thrusting folds, the middle complex of synclines goes from Urbania to Visso and it's composed by east-vergent thrust sheets, while the external wrinkle ridges is an anticlinal structure thrusting over the foothills, named overthrust of the Sibillini Mountains [29]. Finally, going eastwards, the foothills can be divided into two geomorphologic structures: the "pedeappennino marchigiano", characterised by anticlines with transpressive and normal faults, and the periadritic basin with small folds east-vergent.

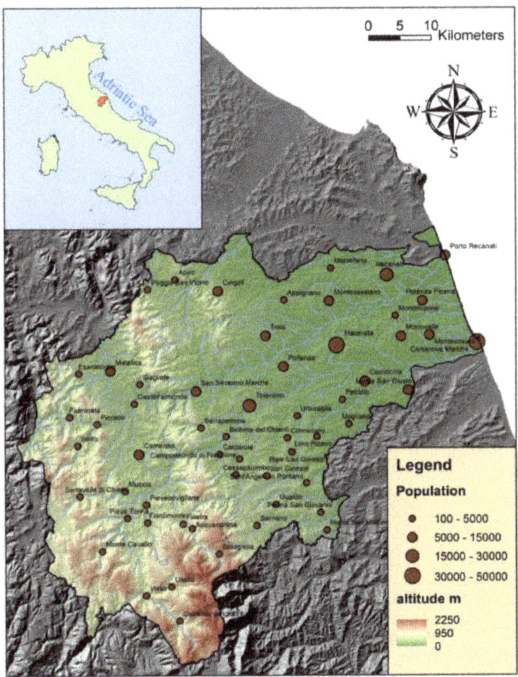

Figure 1. Geography of the province of Macerata [23].

From the point of view of landslides, the province of Macerata is a very heterogeneous territory, with movements of very different types, often grouped by homogeneous zones of acclivity or in relation to the geological substrate. In correspondence of mountain ridges and steep slopes characterised by predominantly calcareous rocks, collapse phenomena and deep-seated gravitational slope deformations (DGSD) are observed. Also in the high energy areas of the relief, there are frequent phenomena of slide, debris flow and debris avalanches, which involve eluvial colluvial deposits and clastic materials accumulated in previous morphoclimatic phases. In the areas with outcrops of Plio-Pleistocene sediments, mainly pelitic, and characterised by a lower gradient, the type of movement that prevails is that of earthflow. Less deep phenomena such as soliflux landslides and plastic deformations are also widespread in these areas. In the impluvial areas, where there are considerable thicknesses of altered and and eluvial deposits, there are frequent mudflows originated during heavy rainfall. In the hilly areas where the Plio-Pleistocene pelitic and pelitic-arenaceous sediments outcrop, the natural instability of these soils has been accelerated by poor land management and, above all, by less maintenance management and, above all, less maintenance of the surface water drainage network. Moreover, the profound changes in the production methods of the agricultural system, which can be summarised as a reduced anthropic presence in the area and a decrease in vegetation cover, have led to the breakdown of delicate natural balances over the last thirty years. The development of settlements and infrastructures, imposed by new socio-economic processes, has often taken place in an uncontrolled manner, occupying areas whose stability was considered precarious.

Moreover, in the last period, this area of Italy has suffered periodically from strong hydrogeological instability, due to two major seismic events in 1997 and 2016, which mainly generated deep-seated gravitational slope deformations (DGSD) and collapses. In addition there have been extreme precipitation events such as the one in November 2013, which activated existing landslides and uncovered new ones, especially in hilly areas.

2. Materials and Methods

2.1. Data Sampling and Preparation

For the analysis of susceptibility through GIS software, a detailed digital elevation model (DEM) is primary, which was created with the help of the regional technical map (CTR) [30]. This DEM was prepared with a resolution of 5m and on this basis the slope map, which is very influential on landslide susceptibility, was obtained. The geological map was digitized and the landslide map was obtained from the "River Basin Authorities of the Marche Region". The model validation was instead produced by introducing the landslides from the IFFI project (inventory of landslide phenomena in Italy). Deep-seated gravitational slope deformations (DGSD) and collapses were excluded from the landslide map, due to activation phenomena not directly linked to extreme precipitation events, thus the total number of landslides considered for this study was 4171 (Figure 2).

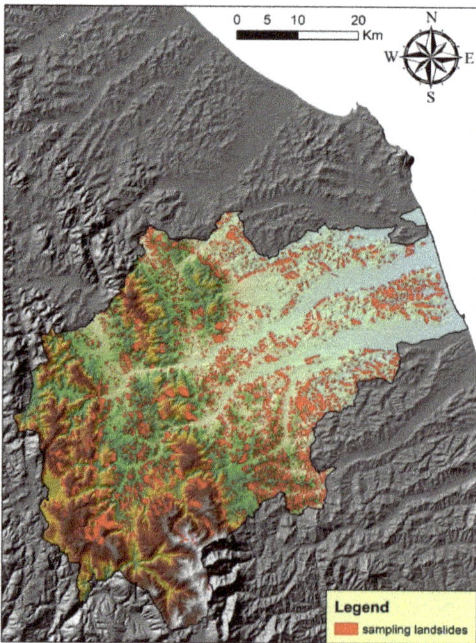

Figure 2. Map of sampled landslides.

The land use map, on the other hand, was obtained from ISPRA (Istituto Superiore per la Protezione e la Ricerca Ambientale), the italian institute that distributes the Corine Land Cover for Italy, developed by Copernicus Global Land Services (CGLS), Europe's leading Earth monitoring programme. In order to complete the parameters that are part of the model, the precipitation of the last 30 years were taken into account, through data of 10 rain gauges in the province of Macerata and another 10 outside. The rainfall data were collected by the Regional Civil Protection of the Marche Region and the Experimental Geophysical Observatory of Macerata (OGSM). Firstly, a complete validation and homogenisation of the climate data was carried out, following the guidelinesof the WMO (World Meteorological Organization). Interpolation was carried out throughout the province by means of ordinary cokriging based on altitude as an independent variable [31]. Ordinary cokriging (OCK), is a geostatistical method used in relation to one or more independent variables [32] that allow a better interpolation if there is a strong correlation between independent variable (known throughout the territory) and dependent one (only some sample values).

$$\begin{aligned} Z_{OCK}(u) = \sum_{\alpha_1=1}^{n_1(u)} \lambda_{\alpha_1}^{OCK}(u) Z_1(u_{\alpha_1}) + \sum_{\alpha_2=1}^{n_2(u)} \lambda_{\alpha_2}^{OCK}(u) Z_1(u_{\alpha_2}) \\ \lambda_{\alpha_1}^{OCK}(u) \text{ and } \lambda_{\alpha_2}^{OCK}(u) &= \text{weights of the data} \\ Z_1(u_{\alpha_1}) \text{ and } Z_1(u_{\alpha_2}) &= \text{primary and secondary data} \end{aligned} \tag{1}$$

The altitude was chosen as an independent variable on the basis of a previous study showing that it is the most correlated topographical parameter for this area [33]. Furthermore, a complex study was performed to find out the amount of precipitation in case of extreme events. The method used to carry out the analysis was the Generalized Extreme Value (GEV), chosen after an assessment of the goodness of fit in relation to precipitation data. The GEV is a flexible model composed of three parameters: k for shape, σ for scale and μ for location.

$$f(x) = \begin{cases} \frac{1}{\sigma} \exp(-(1+kz)^{-1/k}(1+kz)^{-1-1/k} & k \neq 0 \\ \frac{1}{\sigma} \exp(-z - \exp(-z)) & k = 0 \end{cases} \tag{2}$$

where $z = \frac{(x-\mu)}{\sigma}$

The domain of the GEV depends on k:

$$\begin{aligned} 1 + k\frac{(x-\mu)}{\sigma} > 0 \quad k \neq 0 \\ -\infty < x < +\infty \quad k = 0 \end{aligned} \tag{3}$$

In order to assess the goodness of fit for each rain gauge, it was used the R software with the package "extremes 2.0" analyzing the quantile plot and the histogram of frequency [34]. Even the same software was used to calculate the return period. In fact the return period $1/p$ was obtained through the procedure of the maximum likelihood z_p with a chance between 0 and 1:

$$z_p = \mu + \frac{\sigma}{k}\left([-\log(1-p)]^{-k} - 1\right) \tag{4}$$

Finally it was calculate the confidence interval of each return period in this way:

$$\mu = z_p + \frac{\sigma}{k}\left\{1 - [-\log(1-p)]^{-k}\right\} \tag{5}$$

However, although the altitude is optimally correlated with the rainfall, it is not at all correlated with the extreme rainfall events. Thus to have a good reliability, the rain gauges of extreme events in 24 h near to the location of the analysis were interpolated with an ordinary kriging (without altitude), instead of OCK. Ordinary kriging (OK) uses a semivariogram to express the strength of the spatial correlation as a function of distance and similarity.

$$Z_{OK}(u) = \mu + \varepsilon(u) \tag{6}$$

μ = unknown constant, $\varepsilon(u)$ = random error

The goodness of interpolations was evaluated with a cross-validation, performed with GIS softwares, considering some statistical operators as: Mean Error, Root Mean Square Error, Average Standard Error, Mean Standardized Error and Root Mean Square Error Standardized [23]. With regard to extreme climatic events, the analysis was conducted on a return time of 100 years for extreme climatic events considering time series of 50–60 years of precipitation data for the hours 1-3-6-12-24 (Table 1). The confidence interval were calculated through the "bootstrap" method, with 1000 attempt.

Table 1. Example of calculation of return period 100 years of precipitation for Tolentino rain gauge.

Rain Gauge	Return Period 100 Years (mm)
Tolentino 1 h	58.0
Tolentino 3 h	72.3
Tolentino 6 h	84.8
Tolentino 12 h	108.8
Tolentino 24 h	137.9

The results of the analysis are showed with the Extreme Rainfall Intesity-Duration-Frequency (IDF) curve (Figure 3), which relates the precipitation in millimeters to the return period in years.

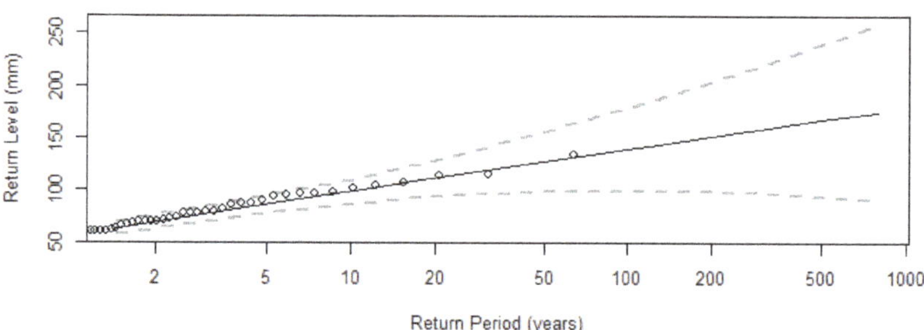

Figure 3. IDF curve of Tolentino for the interval time of 24 h. Dotted line is the confidence interval after 1000 attempts. The black line is the one risulting from the analysis.

2.2. Model Building

Following this in-depth climatic analysis, the most relevant environmental problems were identified, for this territory, according to databases obtained from the Basin Authority of the Marche Region and the Marche Region itself. Landslides detected in the investigation area have been mapped and subsequently combined with the following parameters: extreme events of precipitation, average annual precipitation, geology, land use and slope angle, in order to predict quiescent or potential landslides. The evidences were divided in classes and this analysis was based on the weight of each single class of values. Weight is a function of how many landslides are present in each class and the final aim is to produce a landslide susceptibility map. To create the susceptibility map, the classes of the various evidences climatic interpolations (average precipitation and extreme events), lithology, slope and land use become the subject of the WoE calculation (Figure 4). This calculation performed by means of math tool between raster with GIS software, produces positive and negative weights for each class (Figure 5). Weights are estimated to be proportional to the influence of each class on landslide and were calculated by the following equations [35]:

$$W^{+} = \ln\left(\frac{\frac{\text{Landslide area in calss}}{\text{Total landslide area}}}{\frac{\text{Stable area in class}}{\text{Total stable area}}}\right) \tag{7}$$

$$W^{-} = \ln\left(\frac{\frac{\text{Total landslide area outside calss}}{\text{Total landslide area}}}{\frac{\text{stable area outside class}}{\text{Total stable area}}}\right) \tag{8}$$

The Equations (7) and (8) represents the start of the WoE method, which combine evidence in support of an hypothesis. In this way can be possible to calculate the degree of influence of each factors in the susceptibility analysis, with the aim of produce a map useful to protection. However in this calculation it is essential to know the prior probability (O_f) to find the amount of study area affected by landslide (A_f) over the whole study area (A_t) [20]:

$$O_f = \frac{\frac{A_f}{A_t}}{1 - \frac{A_f}{A_t}} \tag{9}$$

Furthermore there is another very important parameter which is the contrast (C) that represents the differences between W^+ and W^- allowing the assessment if the investigated factor is significant and influence the distribution of landslides in the area. A value of "C" close to 0 determines that the parameter is of little significance, while a value of 2 attests a good correlation. The final susceptibility map was obtained from the weights of each parameter and the prior probability [20]:

$$\text{Final P.} = \text{EXP}\left(\sum W^{+} + \ln O_f\right) \tag{10}$$

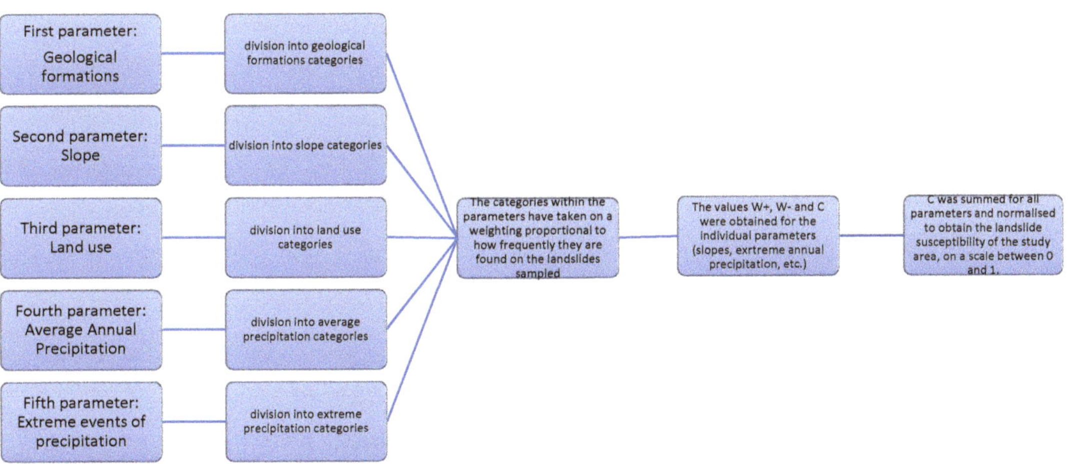

Figure 4. Model flow diagram.

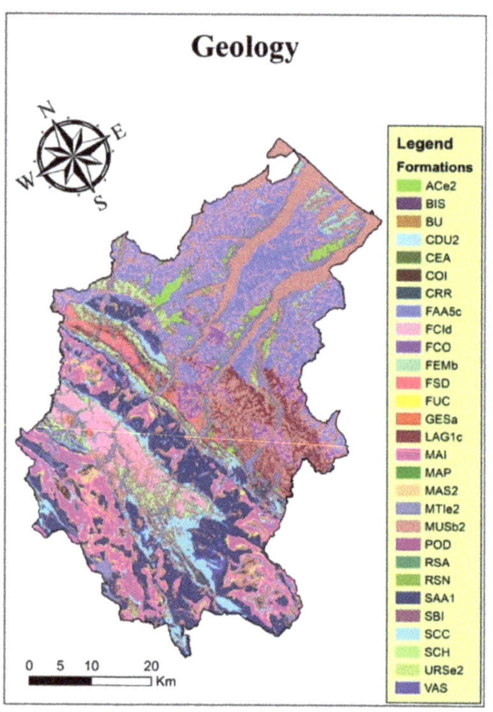

Geological Formation	W+	W-	C
Depositi Continentali Quaternari (MUS)	0.61	-0.47	1.09
Rosso Ammonitico (RSA)	0.29	0.00	0.29
Formazione a Colombacci (FCO)	0.19	-0.00	0.19
Formazione di San Donato (FSD)	0.18	-0.00	0.19
Formazione delle Argille Azzurre (FAA)	0.18	-0.03	0.21
Formazione della Laga (LAG)	0.15	-0.01	0.16
Schlier (SCH)	-0.13	0.00	-0.13
Formazione di Camerino (FCI)	-0.14	0.00	-0.14
Formazione Gessoso-Solfifera (GES)	-0.26	0.00	-0.26
Marne di Cella (CEA)	-0.44	0.00	-0.44
Scaglia Cinerea (SCC)	-0.47	0.02	-0.48
Corniola (COI)	-0.58	0.00	-0.57
Bisciaro (BIS)	-0.60	0.00	-0.60
Marne con Cerrogna (CRR)	-0.64	0.00	-0.64
Marne a Fucoidi (FUC)	-0.71	0.01	-0.71
Scaglia Rossa (SAA)	-0.87	0.09	-0.95
Depositi alluvionali (URSbn)	-0.87	0.00	-0.87
Marne di Monte Serrone (RSN)	-0.87	0.00	-0.87
Formazione di Fermo (FEM)	-0.99	0.00	-1.00
Calcare Massiccio (MAS)	-0.99	0.01	-1.00
Sintema di Matelica (MTI)	-1.14	0.07	-1.20
Scaglia Bianca (SBI)	-1.21	0.01	-1.23
Maiolica (MAI)	-1.26	0.06	-1.32
Calcari a Posidonia (POD)	-1.30	0.00	-1.30
Scaglia Variegata (VAS)	-1.46	0.01	-1.46
Alluvioni Terrazzate (AC)	-1.54	0.02	-1.56
Gruppo del Bugarone (BU)	-1.64	0.00	-1.64
Calcari diasprigni (CDU)	-1.75	0.01	-1.76
Marne a Pteropodi (MAP)	-3.01	0.00	-3.01

Figure 5. (**Left**) Geological map of Macerata Province; (**Right**) The weights calculated for the geology parameter, (W^+) positive weights, (W^-) negative weights, (C) contrast; formations are described by the CARG PROJECT [36].

Geology and Land use are categorized variable, therefore they did not need to be categorized. On the other hand, choices were made for both climatic parameters and slope gradients. Extreme precipitation events were divided into intervals of 5 mm of precipitation while annual precipitation was divided into intervals of 150 mm of precipitation. The slopes were divided into four different classes, the first for assessing flat surfaces, the second for assessing medium-low slopes, the third for medium-high slopes and the fourth for high slopes. Obviously, these subdivisions are arbitrary and could influence the results of the model to a greater or lesser degree. The only way to assess the presence of more appropriate categories, would be to iteratively evaluate them.

3. Results

The landslide map (Figure 2) was overlapped with each influencing parameter in order to find a statistical correlation. The weight of each parameter is a function of the correlated density of instability. The sum of the different parameters determines a landslide susceptibility map. The various thematic maps were overlapped with the landslide map and the intersections obtained with GIS software, were assessed to calculate the weights and the odds for the whole Province of Macerata. The WoE were obtained from 5 parameters Geology (Figure 5), Slope (Figure 6), Land use (Figure 7), Annual average precipitation (Figure 8), Extreme events (Figure 9).

It is important for the Figures 5–9 (right) to observe the contrast ("C") value, because a positive one determine that landslides occur more frequently in the given class. For geology (Figure 5) we have an high value of C for "Depositi Quaternari" (Quaternary deposits) and positive but lower for RSA, FCO, FSD, FAA and LAG [36]. All the other formations do not have a positive correlation of parameter C, which suggests that landslides do not occur very frequently in these geological formations.

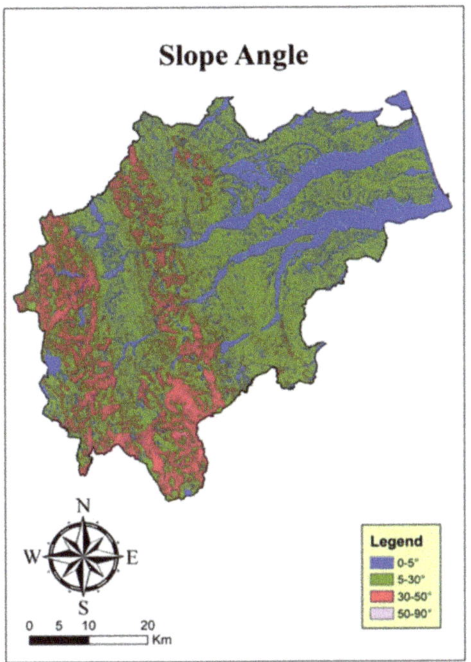

Slope classes	W+	W-	C
0°–5°	-2.778	0.122	-2.900
5°–30°	1.241	-0.839	2.080
30°–50°	-2.938	-0.134	-2.804
50°–90°	-6.428	-0.084	-6.344

Figure 6. (**Left**) Map of slope angle of Macerata province; (**Right**) The weights calculated for the slope parameter.

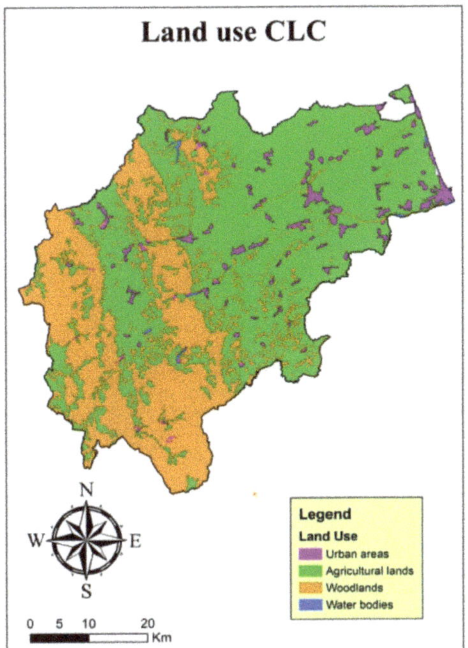

Land use	W+	W-	C
Agricultural lands	0.485	-0.605	1.090
woodlands	-1.127	0.014	-1.140
Urban areas	-4.857	-0.140	-4.717
Water bodies	-7.704	-0.158	-7.546

Figure 7. (**Left**) Land use of Macerata province, from Corinne Land Cover; (**Right**) The weights calculated for the land use parameter.

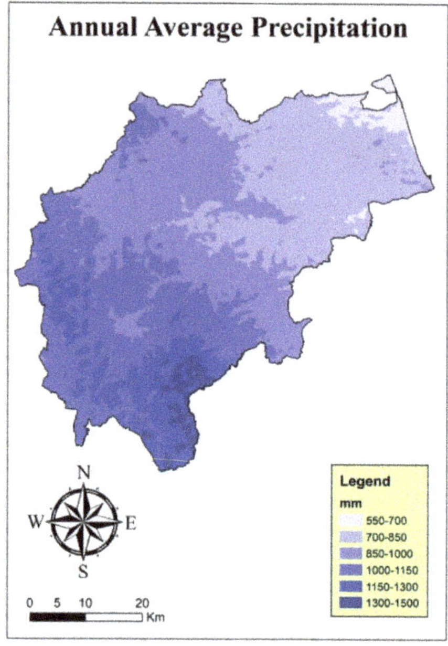

Annual average precipitation (mm)	W+	W-	C
550-700	-4.665	-0.146	-4.519
700-850	-1.006	-0.197	-0.808
850-1000	-0.817	-0.267	-0.550
1000-1150	-1.235	-0.133	-1.102
1150-1300	-2.846	-0.111	-2.735
1300-1500	-5.463	-0.140	-5.323

Figure 8. (**Left**) Annual average precipitation in Macerata Province; (**Right**) The weights calculated for annual average precipitation parameter.

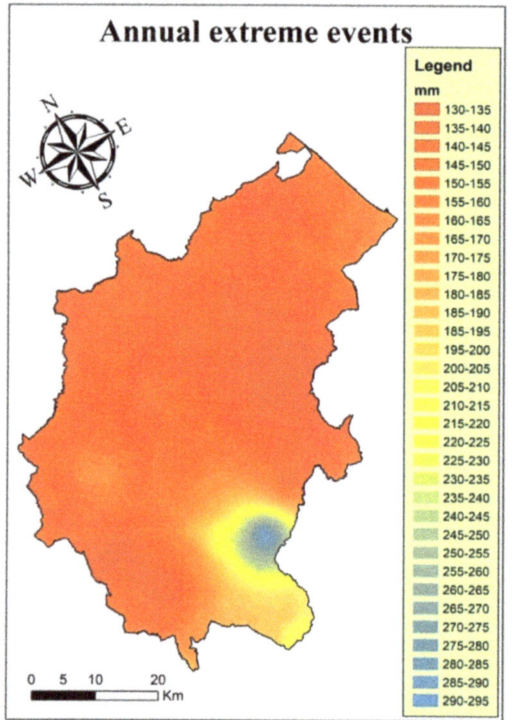

Annual extreme events mm	W+	W-	C
130-135	-2.564	-0.204	-2.360
135-140	-2.122	-0.182	-1.940
140-145	-1.104	-0.183	-0.921
145-150	-1.894	-0.153	-1.741
150-155	-2.204	-0.169	-2.035
155-160	-2.331	-0.161	-2.171
160-165	-3.329	-0.158	-3.171
165-170	-4.066	-0.157	-3.909
170-175	-4.607	-0.159	-4.447
175-180	-4.697	-0.162	-4.534
180-185	-4.864	-0.161	-4.703
185-190	-5.034	-0.157	-4.877
190-195	-5.017	-0.153	-4.864
195-200	-5.864	-0.154	-5.711
200-205	-5.681	-0.154	-5.526
205-210	-5.859	-0.155	-5.704
210-215	-6.377	-0.158	-6.220
215-220	-5.836	-0.161	-5.675
220-225	-6.661	-0.160	-6.501
225-230	-6.536	-0.161	-6.375
230-235	-6.417	-0.161	-6.256
235-240	-6.692	-0.161	-6.532
240-245	-6.609	-0.161	-6.448
245-250	-7.404	-0.160	-7.243
250-255	-8.108	-0.160	-7.948
255-260	-7.971	-0.160	-7.811
260-265	-8.149	-0.161	-7.988
265-270	-12.144	-0.161	-11.984
270-275	-8.557	-0.161	-8.396
275-280	-7.286	-0.162	-7.124
280-285	-7.537	-0.162	-7.375
285-290	-6.448	-0.163	-6.285
290-295	-8.015	-0.162	-7.853

Figure 9. (**Left**) Annual extreme events of the Macerata Province; (**Right**) The weights calculated for extreme events parameter.

The most frequent class of lanslides for the slope angle is between 5° and 30°, while C for all the other classes seems to be not very significant (Figure 6).

The land use (Figure 7 shows a higher contrast value for territories used for agricultural practice as expected. In fact, from the table (Figure 7) seems the agricultural working of the soil exposes it to problems of instability.

Average annual precipitation (Figure 8) not seem to be an highly correlated parameter, and the most influent can be considered for the band 850–1000 m a.s.l.. It is interesting to note a sort of inverse correlation between the amount of precipitation and the contrast, perhaps distorted by the presence of lithologies less susceptible to landslides.

Average annual precipitation and extreme events (Figure 9) do not show values that are decisive for the assessment of the landslide susceptibility, even if there are classes with higher values of contrast than others. In any case, a strong relationship between extreme precipitation and landslide susceptibility has not been found, which even highlighted areas with low extreme precipitation as the most susceptible.

At the end of this procedure, all the results have been overlapped in order to create a landslide susceptibility map; the value as specified in the methods was calculated on the basis of the Equation (10). The weight of evidence for the province of Macerata is represented by the map (Figure 10) in 5 levels of landslide susceptibility from S1 to S5 with each corresponding to a value between 0.0 and 1.0. Territories with a low probability of being affected by landslides were classified as S1 and S2, a result from S3 to S4 has a landslide susceptibility that starts to become important, while level S5 is an area subject to major hydrogeological instability.

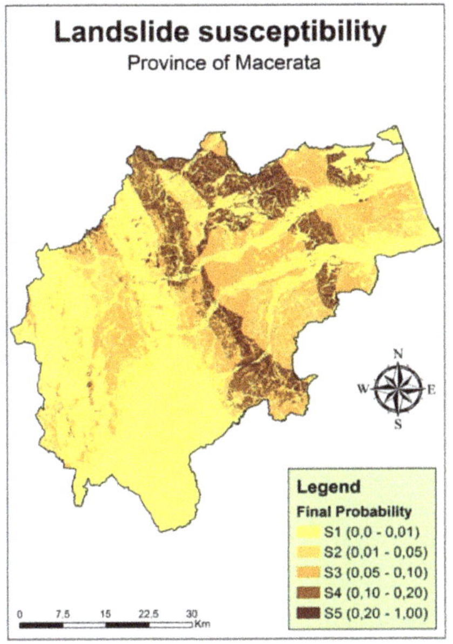

SUSCEPTIBILITY LEVEL	AREA km²
S1	522.746
S2	335.212
S3	447.332
S4	205.029
S5	133.860

Figure 10. (**Left**) Landslide susceptibility of the Province of Macerata; (**Right**) Area of landslides introduced as a result of the validation procedure, in relation to the landslide susceptibility level predicted by the model of this study.

It is important to note in the figure (Figure 10) that in the South-western area, the most mountainous one, the level of landslide susceptibility is unusually low, due to the weight geological formations which have a very low contrast (C), because of more coherent rocks, less prone to the investigated movements. In fact, as specified in the description of the most common movements for each zone of the study area, the mountainous zone shows deep-seated gravitational slope deformations (DGSD) and collapses, movements not analysed in this study. Similarly, the piedmont area has a susceptibility mainly between level 2 and level 3, which makes it an area of low criticality, although there are some areas where susceptibility reaches level 4–5 and therefore need to be managed appropriately. The riverbeds of the most important rivers obviously have a minimal susceptibility level due to the almost non-existent slope as do the coastal stretches in the eastern part. On the other hand, the territories where susceptibility is between S4 and S5 are located in the hilly part, in the centre part of the study area. These areas show pelitic or pelitic-arenaceous geological formations and a land use more oriented to agriculture which generate the most important instability conditions.

Finally, in order to validate the work, many landslides from the IFFI database that were not sampled to develop the model were included, amounting to a total landslide area of 1644.359 Km^2. This validation led to the assessment that about 70% of the landslides introduced, are located in a territory with susceptibility level from 2 to 5, while only 30% are located in a territory with low landslide susceptibility, so they can be considered not predicted by the model.

4. Discussion

This study is an example of WoE for the creation of landslide susceptibility maps through the use of GIS softwares, with the addition of an accurate analysis of extreme precipitation. Extreme precipitation seems to have, in the literature, a great influence in the territory subject to slow-motion landslides [37,38], because this type of landslides are sensitive to soil saturation conditions. However, in this case, no statistically significant and systematic values of influence of extreme events or average precipitation on landslides were found. In particular, it is interesting to note that the C is greater in a low range of extreme events, such as 140–145, however further clustering of the variable should be tested in order to exclude this parameter from those influential for landslide susceptibility. There is no growth in C to the increase of precipitation, which is also a result of the geographical and geological characteristics of the area. This can lead to the assessment that the extreme events in the area are not so different that they become significant and can discriminate one area from another. The division into too many classes can be influential, but even reduced classes do not have much higher C values. In any case, a significance of the extreme event cannot be excluded, which is widely documented in the case of surface gravitational phenomena [39,40]. Among the discriminating and statistically significant parameters for the production of landslide movements, there are the slope gradient, which from 5 to 30° shows an excellent correlation, the agricultural terrain and the geological formation MUS according to the relevant scientific literature [9,41,42]. The validation procedure allowed the reliability of the model to be assessed at about 70%, in line with many other studies that used the same or different calculation methods [15,43]. Despite the apparent lack of significance of extreme events, the result was nevertheless achieved, in fact a reliable susceptibility map was created according to all the factors considered, which provides a priority for risk mitigation interventions.

5. Conclusions

This outcome, combined with the different parameters mentioned above (geology, slope angle, land use, average precipitation, extreme events), composes a model that leads to an automatic detection of possible landslide areas, in this case very focused on the movements that can be originated by heavy rainfall. It would be interesting to study this area further to evaluate other parameters to be included, in order to take into account all possible landslide movements, without discriminating against some of them. This consideration is very important because it allows to obtain a susceptibility map even where the movements are not clear or studied, but only on the basis of possible combinations. The susceptibility map of the province of Macerata, therefore, can lead to the use of this tool for many protection purposes. This tool could support technical decisions, in order to prioritise interventions in a scientific way. The assessment is currently carried out on the basis of previous evidence or emergencies. In this way it would be possible to prevent the emergencies, improving this map also with other important features like soil type, vegetation cover, etc. Obviously in the future it would be important to support this susceptibility map, with a landslide hazard map, in order to create a real operating system. Then the last step could be to create a risk map that takes into account people, heritage, buildings, but also valuable crops, making a detailed assessment of the stability model.

Author Contributions: Conceptualization, M.G., M.M., G.P.; methodology, M.G.; software, M.G., M.M.; validation, M.G.; formal analysis, M.G., G.P.; investigation, M.G., M.M., G.P.; data curation, M.G.; writing—original draft preparation, M.G., M.M.; writing—review and editing, M.M.; supervision, G.P. All authors have read and agreed to the published version of the manuscript.

Funding: This research received no external funding.

Institutional Review Board Statement: Not applicable.

Informed Consent Statement: Not applicable.

Data Availability Statement: Publicly available datasets were analyzed in this study. This data can be found here: Regione Marche—Sistema Informativo Regionale Meteo-Idro-Pluviometrico; Cartografia on line—IFFI (isprambiente.it).

Conflicts of Interest: The authors declare no conflict of interest.

References

1. Trigila, A.; Iadanza, C.; Bussettini, M.; Lastoria, B. *Dissesto Idrogeologico in Italia: Pericolosità e Indicatori di Rischio—Edizione;* Rapporti 287/2018; ISPRAL: Rome, Italy, 2018.
2. Gariano, S.L.; Guzzetti, F. Evaluating the Impact of Climate Change on Landslide Occurrence, Hazard, and Risk: From Global to Regional Scale. In Proceedings of the 19th EGU General Assembly Conference Abstracts, Austria, Vienna, 23–28 April 2017; Volume 19, p. 503.
3. Winter, M.G.; Shearer, B. Climate change and landslide hazard and risk in Scotland. *Eng. Geol. Soc. Territ.* **2015**, *1*, 411–414.
4. Crowley, T.J.; Berner, R.A. CO_2 and climate change. *Science* **2001**, *292*, 870–872. [CrossRef] [PubMed]
5. Falcinelli, S. Fuel production from waste CO_2 using renewable energies. *Catal Today* **2020**, *348*, 95–101. [CrossRef]
6. Ciabatta, L.; Camici, S.; Brocca, L.; Ponziani, F.; Stelluti, M.; Berni, N.; Moramarco, T. Assessing the impact of climate-change scenarios on landslide occurrence in Umbria Region, Italy. *J. Hydrol.* **2016**, *541*, 285–295. [CrossRef]
7. Gariano, S.L.; Petrucci, O.; Rianna, G.; Santini, M.; Guzzetti, F. Impacts of past and future land changes on landslides in southern Italy. *REG Environ. Chang.* **2018**, *18*, 437–449. [CrossRef]
8. Tasser, E.; Mader, M.; Tappeiner, U. Effects of land use in alpine grasslands on the probability of landslides. *Basic Appl. Ecol.* **2003**, *4*, 271–280. [CrossRef]
9. Chowdhury, R.; Flentje, P. Role of slope reliability analysis in landslide risk management. *B Eng. Geol. Environ.* **2003**, *62*, 41–46. [CrossRef]
10. Pachauri, A.K.; Pant, M. Landslide hazard mapping based on geological attributes. *Eng. Geol.* **1992**, *32*, 81–100. [CrossRef]
11. Yin, Y.; Wang, F.; Sun, P. Landslide hazards triggered by the 2008 Wenchuan earthquake, Sichuan, China. *Landslides* **2009**, *6*, 139–152. [CrossRef]
12. Yilmaz, I. Landslide susceptibility mapping using frequency ratio, logistic regression, artificial neural networks and their comparison: A case study from Kat landslides (Tokat—Turkey). *Comput. Geosci.* **2009**, *35*, 1125–1138. [CrossRef]
13. Lee, S.; Min, K. Statistical analysis of landslide susceptibility at Yongin, Korea. *Environ. Geol.* **2001**, *40*, 1095–1113. [CrossRef]
14. Pourghasemi, H.R.; Pradhan, B.; Gokceoglu, C.; Mohammadi, M.; Moradi, H.R. Application of weights-of-evidence and certainty factor models and their comparison in landslide susceptibility mapping at Haraz watershed, Iran. *ARAB J. Geosci.* **2013**, *6*, 2351–2365. [CrossRef]
15. Vakhshoori, V.; Zare, M. Landslide susceptibility mapping by comparing weight of evidence, fuzzy logic, and frequency ratio methods. *Geomat. Nat. Hazards Risk* **2016**, *7*, 1731–1752. [CrossRef]
16. Bonham-Carter, G.F. Integration of Geoscientific Data Using GIS. In *Geographic Information Systems: Principles and Applications: Longman*; Maguire, D.J., Goodchild, M.F., Rhind, D.W., Eds.; Longdom: Brussels, Belgium, 1991; Volume 2, pp. 171–184.
17. Agterberg, F.P.; Bonham-Carter, G.F.; Cheng, Q.; Wright, D.F. Weights of Evidence Modeling and Weighted Logistic Regression in Mineral Potential Mapping. In *Computers in Geology*; Davis, J.C., Herzfeld, U.C., Eds.; Oxford University Press: New York, NY, USA, 1993; pp. 13–32.
18. Regmi, N.R.; Giardino, J.R.; Vitek, J.D. Modeling susceptibility to landslides using the weight of evidence approach: Western Colorado, USA. *Geomorphology* **2010**, *115*, 172–187. [CrossRef]

19. Kayastha, P.; Dhital, M.R.; De Smedt, F. Landslide susceptibility mapping using the weight of evidence method in the Tinau watershed, Nepal. *Nat. Hazards* **2012**, *63*, 479–498. [CrossRef]
20. Barbieri, G.; Cambuli, P. The Weight of Evidence Statistical Method in Landslide Susceptibility Mapping of the Rio Pardu Valley (Sardinia, Italy). In Proceedings of the 18th World IMACS Congress and MODSIM09 International Congress on Modelling and Simulation: Interfacing Modelling and Simulation with Mathematical and Computational Sciences, Cairns, Australia, 13–17 July 2009; pp. 2658–2664.
21. Piacentini, D.; Troiani, F.; Soldati, M.; Notarnicola, C.; Savelli, D.; Schneiderbauer, S.; Strada, C. Statistical analysis for assessing shallow-landslide susceptibility in South Tyrol (south-eastern Alps, Italy). *Geomorphology* **2012**, *151*, 196–206. [CrossRef]
22. Cervi, F.; Berti, M.; Borgatti, L.; Ronchetti, F.; Manenti, F.; Corsini, A. Comparing predictive capability of statistical and deterministic methods for landslide susceptibility mapping: A case study in the northern Apennines (Reggio Emilia Province, Italy). *Landslides* **2010**, *7*, 433–444. [CrossRef]
23. Gentilucci, M.; Barbieri, M.; Burt, P. Climatic Variations in Macerata Province (Central Italy). *Water Sui* **2018**, *10*, 1104. [CrossRef]
24. Gentilucci, M.; Bisci, C.; Burt, P.; Fazzini, M.; Vaccaro, C. Interpolation of Rainfall through Polynomial Regression in the Marche Region (Central Italy). In *The Annual International Conference on Geographic Information Science*; Springer: Cham, Switzerland, 2018; pp. 55–73.
25. Gentilucci, M.; Materazzi, M.; Pambianchi, G.; Burt, P.; Guerriero, G. Assessment of Variations in the Temperature-Rainfall Trend in the Province of Macerata (Central Italy), Comparing the Last Three Climatological Standard Normals (1961–1990; 1971–2000; 1981–2010) for Biosustainability Studies. *Environ. Process.* **2019**, *6*, 1–22. [CrossRef]
26. Gentilucci, M.; Barbieri, M.; Lee, H.S.; Zardi, D. Analysis of Rainfall Trends and Extreme Precipitation in the Middle Adriatic Side, Marche Region (Central Italy). *Water* **2019**, *11*, 1948. [CrossRef]
27. Beniston, M.; Stephenson, D.B.; Christensen, O.B.; Ferro, C.A.; Frei, C.; Goyette, S.; Halsnaes, K.; Holt, T.; Jylhä, K.; Koffi, B.; et al. Future extreme events in European climate: An exploration of regional climate model projections. *Clim. Chang.* **2007**, *81*, 71–95. [CrossRef]
28. Gentilucci, M.; Barbieri, M.; Burt, P. Climate and Territorial Suitability for the Vineyards Developed Using GIS Techniques. In *Exploring the Nexus of Geoecology, Geography, Geoarcheology and Geotourism: Advances and Applications for Sustainable Development in Environmental Sciences and Agroforestry Research*; Springer: Cham, Switzerland, 2019; pp. 11–13.
29. Dramis, F.; Farabollini, P.; Gentili, B.; Pambianchi, G. Neotectonics and Large-Scale Gravitational Phenomena in the Umbria–Marche Apennines, Italy. In Proceedings of the Seismically Induced Ground Ruptures and Large Scale Mass Movements. Field Excursion and Meeting, Apennines, Italy, 21–27 September 2001; Volume 4, pp. 17–30.
30. Kweon, I.S.; Kanade, T. Extracting topographic terrain features from elevation maps. *Cvgip Image Underst.* **1994**, *59*, 171–182. [CrossRef]
31. Ninyerola, M.; Pons, X.; Roure, J.M. A methodological approach of climatological modelling of air temperature and precipitation through gis techniques. *Int. J. Clim.* **2000**, *20*, 1823–1841. [CrossRef]
32. Qin, Q.; Wang, H.; Lei, X.; Li, X.; Xie, Y.; Zheng, Y. Spatial variability in the amount of forest litter at the local scale in northeastern China: Kriging and cokriging approaches to interpolation. *Ecol. Evol.* **2020**, *10*, 778–790. [CrossRef]
33. Gentilucci, M.; Barbieri, M.; Burt, P.; D'Aprile, F. Preliminary data validation and reconstruction of temperature and precipitation in Central Italy. *Geosciences* **2018**, *8*, 202. [CrossRef]
34. Gilleland, E.; Katz, R.W. extRemes 2.0: An extreme value analysis package in R. *J. Stat. Softw.* **2016**, *72*, 1–39. [CrossRef]
35. Sumaryono, D.M.; Sulaksana, N.; DasaTriana, Y. Weights of Evidence Method for Landslide Susceptibility Mapping in Tandikek and Damar Bancah, West Sumatra, Indonesia. *Int. J. Sci. Res. (IJSR)* **2015**, *4*, 1283–1290.
36. Mattioli, M.; Pieruccini, P.; Pennacchioni, E.; Piergiovanni, A.; Sandroni, P.; Tosti, S.; Tramontana, M. PROGETTO CARG, Note Illustrative Della Carta Geologica d'Italia 1:50.000. 2002. Available online: https://www.facebook.com/PaginaUfficialeRegioneMarche/ (accessed on 1 November 2020).
37. Macfarlane, D.F. Observations and predictions of the behaviour of large, slow-moving landslides in schist, Clyde Dam reservoir, New Zealand. *Eng. Geol.* **2009**, *109*, 5–15. [CrossRef]
38. Ferlisi, S.; Peduto, D.; Gullà, G.; Nicodemo, G.; Borrelli, L.; Fornaro, G. The Use of DInSAR Data for the Analysis of Building Damage Induced by Slow-Moving Landslides. In *Engineering Geology for Society and Territory-Volume 2*; Springer: Cham, Switzerland, 2015; pp. 1835–1839.
39. Clarke, M.L.; Rendell, H.M. Process–form relationships in Southern Italian badlands: Erosion rates and implications for landform evolution. Earth Surface Processes and Landforms. *J. Br. Geomorphol. Res. Group* **2006**, *31*, 15–29.
40. Papathoma-Koehle, M.; Keiler, M.; Totschnig, R.; Glade, T. Improvement of vulnerability curves using data from extreme events: Debris flow event in South Tyrol. *Nat. Hazards* **2012**, *64*, 2083–2105. [CrossRef]
41. Bordoni, M.; Galanti, Y.; Bartelletti, C.; Persichillo, M.G.; Barsanti, M.; Giannecchini, R.; Meisina, C. The influence of the inventory on the determination of the rainfall-induced shallow landslides susceptibility using generalized additive models. *CATENA* **2020**, *193*, 104630. [CrossRef]
42. Pánek, T.; Brázdil, R.; Klimeš, J.; Smolková, V.; Hradecký, J.; Zahradníček, P. Rainfall-induced landslide event of May 2010 in the eastern part of the Czech Republic. *Landslides* **2011**, *8*, 507–516. [CrossRef]
43. Polykretis, C.; Chalkias, C. Comparison and evaluation of landslide susceptibility maps obtained from weight of evidence, logistic regression, and artificial neural network models. *Nat. Hazards* **2018**, *93*, 249–274. [CrossRef]

Article

Characteristics of Rainfall Events Triggering Landslides in Two Climatologically Different Areas: Southern Ecuador and Southern Spain

José Antonio Palenzuela Baena [1,*], John Soto Luzuriaga [2] and Clemente Irigaray Fernández [1]

1 Department of Civil Engineering, University of Granada, 18017 Granada, Spain; clemente@ugr.es
2 Department of Geology and Mining and Civil Engineering, Universidad Técnica Particular de Loja, San Cayetano Alto s/n, Loja 1101608, Ecuador; jesoto@utpl.edu.ec
* Correspondence: jpalbae@ugr.es

Received: 22 June 2020; Accepted: 18 July 2020; Published: 21 July 2020

Abstract: In the research field on landslide hazard assessment for natural risk prediction and mitigation, it is necessary to know the characteristics of the triggering factors, such as rainfall and earthquakes, as well as possible. This work aims to generate and compare the basic information on rainfall events triggering landslides in two areas with different climate and geological settings: the Loja Basin in southern Ecuador and the southern part of the province of Granada in Spain. In addition, this paper gives preliminary insights on the correlation between these rainfall events and major climate cycles affecting each of these study areas. To achieve these objectives, the information on previous studies on these areas was compiled and supplemented to obtain and compare Critical Rainfall Threshold (CRT). Additionally, a seven-month series of accumulated rainfall and mean climate indices were calculated from daily rainfall and monthly climate, respectively. This enabled the correlation between both rainfall and climate cycles. For both study areas, the CRT functions were fitted including the confidence and prediction bounds, and their statistical significance was also assessed. However, to overcome the major difficulties to characterize each landslide event, the rainfall events associated with every landslide are deduced from the spikes showing uncommon return periods cumulative rainfall. Thus, the method used, which has been developed by the authors in previous research, avoids the need to preselect specific rainfall durations for each type of landslide. The information extracted from the findings of this work show that for the wetter area of Ecuador, CRT presents a lower scale factor indicating that lower values of accumulated rainfall are needed to trigger a landslide in this area. This is most likely attributed to the high soil saturation. The separate analysis of the landslide types in the case of southern Granada show very low statistical significance for translational slides, as a low number of data could be identified. However, better fit was obtained for rock falls, complex slides, and the global fit considering all landslide types with R^2 values close to one. In the case of the Loja Basin, the ENSO (El Niño Southern Oscillation) cycle shows a moderate positive correlation with accumulated rainfall in the wettest period, while for the case of the south of the province of Granada, a positive correlation was found between the NAO (North Atlantic Oscillation) and the WeMO (Western Mediterranean Oscillation) climate time series and the accumulated rainfall. This correlation is highlighted when the aggregation (NAO + WeMO) of both climate indices is considered, reaching a Pearson coefficient of −0.55, and exceeding the average of the negative values of this combined index with significant rates in the hydrological years showing a higher number of documented landslides.

Keywords: Critical Rainfall Thresholds; climate cycles; triggering factor; correlation

---

1. Introduction

Landslides are considered to be one of the most serious hydrological hazards, producing isolated or catastrophic events that result in costly damage and high rates of causalities. Quantitative reviews on their effects at different territorial scales can be found in the literature (e.g., [1–6]). These studies show thousands of fatalities a year globally, while the costs caused reach thousands of millions of euros. This phenomenon appears in different types of landslides and under a variety of environmental and climatic scenarios, which make it crucial to understand the hydrological mechanisms that lead to the activation of such land processes with negative effects. In a hazard assessment, it is important to define the thresholds under which landslides develop and their expected frequency. Considering that a threshold is the limit of a quantity from which a process initiates or a state changes [7], a rainfall threshold triggering landslides is related to the limit of hydrological conditions such as soil moisture, rainfall intensity, or accumulated precipitation [8–12]. In the literature, there are numerous investigations on establishing the minimum amount of precipitation that leads to landslides at different locations and scales all over the world. A review of the different types of rainfall thresholds that trigger landslides and the results from different locations are addressed in a more in-depth manner in [10], with corresponding references. To define rainfall threshold, authors use two main types of approaches: physical ones (process-based, conceptual) or empirical ones (historical, statistical). The physical or process-based models incorporate the infiltration pattern into the slope instability analysis. These models are characterized by the difficulty of gathering all the necessary information on hydrological, lithological, morphological, and soil geomechanics parameters. This information gathering is costly and the method is better suited to shallow landslides. The method used in this paper is of the second type, where empirical rainfall thresholds are defined by exploring the rainfall events associated with the occurrence of landslides. Unlike physically-based models, the empirically-based methods for the acquisition phase of data are relatively inexpensive. In fact, the gathering of complementary data, such as piezometric monitoring data and horary rainfall depth, was not available, or the boreoarctic time to be collected for specific sites was too long in comparison with the scarce data that could be expected. Instead, daily rainfall data were provided short-term, and the date of occurrence of the landslides can be extracted from literary sources such as the press, libraries, published scientific papers, and books. The precision of these models depends heavily on the quantity of data available to characterize the hydrological scenarios producing landslides. However, in general, collecting the event dates and detailed rainfall records for long periods is not a simple activity. This is essentially due to the scarce amount of information available, which is neither systematic nor standardized. These data are recorded every time a landslide is witnessed or documented [13,14]. After collecting the necessary data, the empirically-based models are expressed by the correlation of precipitation measurements (accumulated rainfall, duration, and intensity) related to individual or multiple rainfall events leading (or not) to landsliding. When multiple rainfall events are considered, a curve representing the function of the correlation between rainfall measurements can be depicted to visually distinguish between the wet conditions that trigger landslides in an area and the rainfall conditions that do not. This is known as a Critical Rainfall Threshold (CRT) curve. In this paper, the CRTs have been obtained by combining the parameters of rainfall event duration and a measurement of cumulative rainfall. This simple approach, which consists of fitting the function that correlates cumulative precipitation and the duration of the rainfall events that trigger landslides, has been applied by different authors in different ways. Several pieces of research take into account the cumulative event rainfall (E) for shallow soil slips and landslides of a flow-like nature [15–20]. Conversely, deeper-seated landslides need more time for the layered geomaterials involved in the slipping mass to reach predisposing moisture conditions, so the antecedent rainfall (A) as the accumulated precipitation before the landslide event is accounted for [21,22]. Moreover, several analyses combine both cumulative event rainfall and antecedent rainfall [23–26]. Commonly, more than one duration is explored from periods ranging from a few days to several months. However, the selection of the duration interval/s is not so trivial if the information gathered is not precise and characterized by the uncertainty regarding the type, geometry, or depth of the failures being evaluated. This fact makes it difficult to apply a single rule to the landslide events being studied. In this research, the main issues hampering the main objectives coincides with those described in [14]. First, not all the public local organizations have implemented procedures to facilitate timely and affordable access to the necessary information and data, which delays or even prevents the most complete records on hazardous events. Similarly, some private companies are reluctant to provide data companies with their data for reasons of legal protection, data privacy, or other reasons of the company itself. Second, the supplied data and exploited information, such as the information from local newspapers, seldom includes the most relevant data to solve the basic problems involved in a landslide hazard analysis. That is basically the type of landslide, its geometry and dimensions, the description of the mobilized materials, and the hydrogeological or geomechanical parameters. This also applies to landslide or hydrogeological hazard catalogues or databases. Thus, given the incomplete and imprecise nature of the necessary information, some considerations were taken into account. Considering the above-mentioned lack of data and uncertainties about the landslide type, this research applies a method to extract the information on the duration and coupled cumulative rainfall regardless of landslide type, although a separation has been carried out where the information gathered provided the classification of the different landslides. In addition, this method evaluates a duration interval range from 1 to 90 days for each landslide event, with the selection of the most significant rainfall and its respective duration. The selection of the rainfall event is based on the joint visualization of the return periods of the different cumulative rainfalls, which help to distinguish those uncommon events that are more probably related to the triggering of the recorded landslides. Thereby, the selected rainfall events can include, or not include, the cumulative event rainfall related to the date of the landslide event or a few days previous to its initiation, as discussed in [13]. Although, in some cases, the type of landslide cannot be identified or was not correctly collected in the consulted sources, this method has the advantage of being independent of the type of landslide. In other words, the rainfall triggering each landslide is deduced from the spikes coinciding with uncommon return periods of rainfall events with specific accumulated rainfall and duration. This means that the duration of the rainfall events is not predefined by the expert, but is rather derived from each uncommon rainfall event detected from the return period spectrum after considering the wide range of rainfall durations (1–90 days). Accordingly, this method has been applied to both case studies as it can be used for a single type of landslide or for a mixed dataset with different types of mass movements.

In previous research [13,27], the proposed method has been applied by using different functions and rainfall parameters in both study areas. Conversely, this paper seeks to homogenize the CRT analysis by using the same functions for every case, providing a comparison between both scenarios. In addition to the information on triggering conditions, the present research deal with the correlation of these rainfall thresholds and the landslide occurrences with the main climatic cycles affecting the area under analysis. Both the weather models and CRTs (Critical Rainfall Threshold) provide valuable information for the prediction of landslides events, despite the changing patterns in the space and time of climatic cycles.

Considering the significance of this information as a contribution to upcoming research, this work aims to produce initial insights on the CRT functions extracted from the information gathered for two zones that are climatologically and environmentally different. The first zone is located at the Loja Basin in southern Ecuador, while the second one is sited in the south of the province of Granada (Spain). As a second objective, this work adds a graphical and quantitative analysis of the accumulated rainfall during the wettest period and the relevant climatic indices greatly affecting each study area.

2. Insights on the Association of Teleconnections with Floods and Landslides

The teleconnection indices represent the variability in the flows or circulations that favors the appearance of precipitation and controls the patterns of wet and drought periods.

2.1. ENSO (El Niño Southern Oscillation)

The ENSO [28] represents the "balance" of sea pressure in the tropical Pacific Ocean. This phenomenon develops in two phases. The warm phase "El Niño" [29] constitutes the oceanographic and atmospheric periodic anomalies along the Oriental Equatorial Pacific and the southern coastal zone of Ecuador and Peru. This phase causes sea surface temperature (SST) to increase in the Equatorial Pacific. At the same time, SST warming favors evaporation and troposphere heating, leading to atmospheric convection in its warm phase called El Niño. However, the cold phase, La Niña, can cause droughts in multiple areas of South America. A warm phase is declared when the average of the SST anomalies in the Niño 3–4 region (5N–5S, 170W–120W) are above the threshold of +0.5 °C for five consecutive 3-month periods. The intensity of this phenomenon is measured by the Oceanic Niño Index (ONI) as the deviation from this standard (greater than or equal to +0.5 °C). Reference [30] states that the Southwest Pacific (SWP) droughts, landslides, and coastal flooding are linked to ENSO and tropical cyclones. For example, during the moderate phases of El Niño in 2009 and La Niña in 2010, the above average landslide activity was detected peaking in November 2010 in Colombia and Venezuela [31,32]. Also, 54% of the landslides accounted for within the period 2004–2016 in Brazil occurred between 2009 and 2011, with peaks from December 2009 and April 2010 coinciding with a strong El Niño, and January 2011 with a strong La Niña [33]. Similarly, "The Callapa mega-landslide" was developed during the La Niña event of 2010–2011 in Bolivia [34].

In Colombia, La Niña events are associated with a high frequency of flooding, mass movements, and infrastructure damages [35] (and references therein). The review of the database DesInventar for the Antioquia (Colombian Andes) by [36] accounted for 3478 mass movements that caused 2065 deaths and 74654 damaged houses during the period 1900–2017. The study area showed a greater number of landslides during La Niña of almost equal distribution during the hydrological year.

In central-northern Chile, in the mid basin of Elqui River, at 40 km from the Pacific Ocean and an elevation range between 400 and 3900 m, [37] (and references therein) concluded that the ENSO and the number of landslides present a positive correlation.

Moreover, the more eastern countries (closer to the Atlantic Ocean) have been impacted by excessive rainfall during ENSO, causing documented disasters. In other parts of the continent, the ENSO effects do not show significant differences between the number of landslides linked to El Niño, La Niña, and neutral years, probably due to the topographic features and the complex interaction of the Pacific and Atlantic anticyclones [38]. This is well noted in northwestern Argentina to the south of the Cordillera Oriental in the Central Andes [39]. However, the increasing number of landslides have been also recorded in this area during the El Niño warm phases [38] (and references therein). This increasing was found during the extraordinary storms of 1982–1983, 1991–1992, and 2015–2016 when a high number of debris flows and rockfalls caused incalculable damage to the International Road to Chile and the Transandine International Railway on the margins of the Mendoza River [38,40].

When the SST rises due to ENSO events, the coastal northern Peru and southern Ecuador are those which are most directly impacted as the warm phase destabilizes the lower atmosphere [41].

In Peru, the natural hazards are more frequent in coastal areas and in the Cordillera Occidental and Cordillera Central, decreasing as one goes inland [42]. Despite the hyper-arid coastal climate of northern Peru, the authors of [43,44] concluded that the region around the city of Trujillo experiences some of the strongest ENSO signals associated with increases in rainfall, typically exceeding 40%. For example, [44] found regional correlations between 0.58 and 0.78, considering the stations in the cities of Trujillo, Chiclayo, and Piura. Similarly, a review of the homogeneous landslide database DesInventar for the area of the Ancash Department (Peru) showed a major correlation of landslides with El Niño episodes (41% of all the recorded catastrophes) between 1971 and 2009 [42]. More recently, multiple regions located on the coast of northern Peru were seriously affected by floods during the first three months of 2017 with the increasing SST yield in the El Niño of 2017 [45]. In that period, this area received 10 times its average rainfall.

The population and economic activity of the coastal zone of Ecuador are high. Thus, multiple investigations have contributed to the Ecuadorian rainfall prediction in relation to El Niño events that have caused floods and landslides, millions dollars' worth of damage, and thousands of deaths [46] (and references therein). In this area, historical El Niño events identified in 1982–1983, 1987–1988, 1991–1992, and 1997–1998 favored rainfall anomalies with valuable damage to livelihoods, agriculture, and infrastructure.

2.2. NAO (North Atlantic Oscillation)

The NAO is one of the main sources of climate variability in the European region [47–49]. This is formed by the pressure difference between two opposite sign action centers located in the North Atlantic, one over the Azores and another over Iceland [50–53]. The NAO is correlated with temperature [54,55] and precipitation extremes [56]. Several authors have demonstrated that the negative NAO phase has favored anomalously high atmospheric instability over southern Europe, the Mediterranean Basin, northern Africa, and northwest Africa for the last four decades in winter [57] (and references therein).

The NAO climatological phenomenon has notable effects on the northern Iberian Peninsula, usually between December and February. Negative NAO and lower solar activity were primarily responsible for the high frequency of floods during the autumns and winters of the last 2000 years in western-central Europe. This information is derived from sedimentological studies on slackwater flood deposits [58]. The expansion of this phenomenon has also been interpreted to reach the Mediterranean moisture [59]. In the Mediterranean regions, significant storms and high levels of accumulated precipitation are detected due to the changes in the atmospheric winter flow that can last even four months (December–March) in extreme cases. The interaction of climatic influences and the understanding of their trends is complex. Nevertheless, efforts have been made to study the interval between prolonged and intense rainfall episodes resulting from the influence of the NAO. Portugal and Spain are two countries directly affected by the NAO. In Portugal, [60] found significant anti-correlation by applying the Pearson coefficient to the relationship between several precipitation indices and the NAO index, except for the CDD (cumulative dry days).

Using the global extension of teleconnections, the NAO has also had a significant influence on southern Spain, producing a decreasing trend in the annual rainfall in the western Mediterranean and a rising number and intensity of heavy storms of very short duration [61–64]. In the Tramuntana Range at the north-western part of the Majorca Island (central Mediterranean basin), [65] conducted a spectral analysis to determine the presence and statistical significance of climate cycles. Thus, the analysis of different long-term rainfall series permitted the identification of ENSO, NAO, Quasi-Biennial Oscillation (QBO), and Hale and Sun Spot cycles as other signals related to solar activity. Among these, the NAO cycles showed one of the most powerful signals (peaks) in the six-month frequency. In this area, the NAO and ENSO peaks were well-matched with numerous landslide events from a detailed inventory (174 events) dating back to 2005. That is the case of the period 2008–2010, when the island of Majorca experienced the coldest and wettest winters of the last 40 years. Coinciding with the high NAO and ENSO values of that period, 66 mass movements occurred in the Tramuntana range [65,66].

2.3. WeMO (Western Mediterranean Oscillation)

The WeMO index permits the study of rainfall patterns on the Iberian Peninsula, where NAO has a lesser effect [67–69]. The WeMO index (WeMOi) is calculated as the standardized pressure difference between Padua (northern Italy) and San Fernando (southeastern Spain) [67].

The literature on the influence of WeMO is very limited. Nevertheless, the research carried out by [69] analyzed the dependency of rainfall on the WeMO index (WeMOi) in Catalonia (northeastern Spain). They applied the Pearson correlation coefficient to the subperiods 1951–1981 and 1983–2014. The first subperiod showed a significant negative correlation between annual rainfall totals and WeMOi, while in the second interval, this significant correlation disappeared completely. It was also observed that the decrease in WeMO was followed by a decrease in the variability of several rainfall indices, such as the variation coefficient (VC) and the consecutive disparity index (S1), though not for the CI (daily concentration index). Thus, it was demonstrated that WeMO also has the potential to identify the degree of rainfall variability. The WeMO was more sensitive for rainfall series with greater VCs, but had almost null influence in the summer storms when the baric gradient was low [69].

3. Study Areas

3.1. Loja Basin (Southern Ecuador)

The area of the Loja Basin is characterized by a climate classed as warm and fully humid with a warm summer (Cfb) [70] at an average elevation of 2100 m a.s.l., between the meridians 79°10′ and 79°15′, and between the parallels 3°55′ and 4°5′ (Figure 1). The low strength properties of the rocks forming the gentle slopes of the valley, including the city area, appear as a main conditioning factor. This particular geological setting, together with low but very continuous daily rainfall, is the main cause of the frequent occurrence of landslides in this area [27]. This valley presents an average monthly temperature of 16.2 °C, reaching minimums in the coldest month of July (14.9 °C), and its average annual rainfall is 917 mm [71]. The continuity of the rainfall is represented by 59% of the pluviometry records gathering quantities with a mean of 2.56 mm/24 h, and with the rainiest season concentrated from December to April. The histogram depicting the monthly rainfall for the record covering the years ranging from 1964 to 2015 shows maximum convexity during the October–April period (Figure 2). It is in this period when storms and longer precipitation periods trigger floods and mass movements. The weather in the southern part of Ecuador is mainly controlled by the global pressure systems of the subtropical highs over the Atlantic Ocean and the subtropical high over the southeastern Pacific. These systems contribute to the main westward wind streams arriving in southern Ecuador through the mid-level layers (~850 to ~400 hPa) [72].

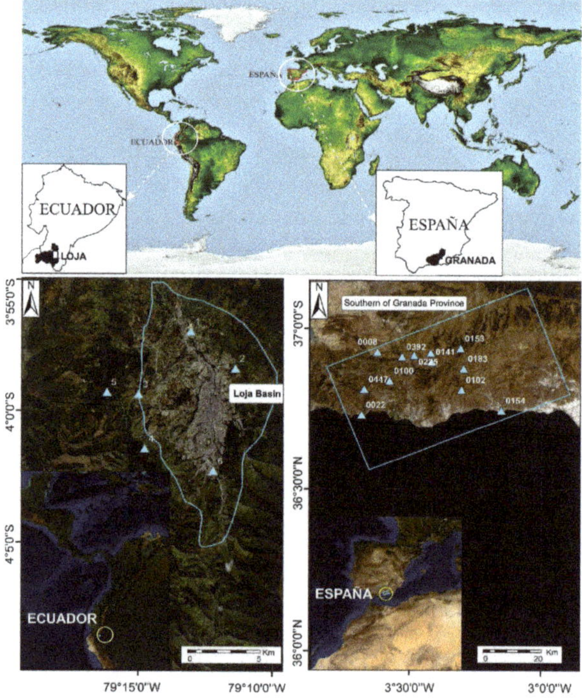

Figure 1. Location of the study areas. Basemap from ArcGIS® World Imagery.

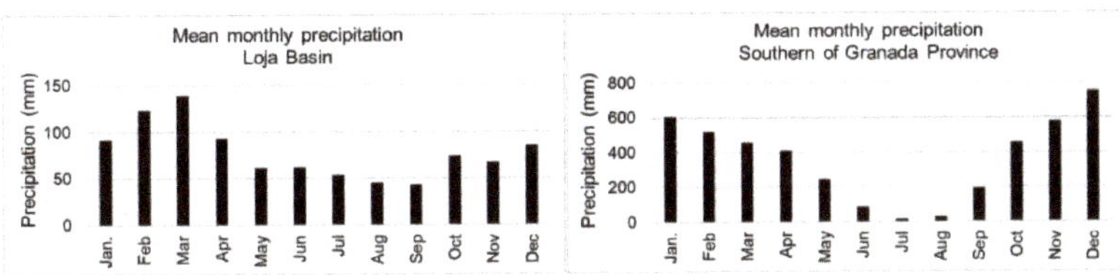

Figure 2. Histogram of the mean monthly precipitation for both study areas.

The Loja Basin is one of several Neogene intramontane basins that have been described in southern Ecuador. Geologically, it is a sedimentary basin of lacustrine origin of the Miocene–Pliocene age, with lithologies such as coarse-grained sandstones with intercalated sediments and conglomerates, thick beds of massive limestone and layers of marl, and fine-grained sandstones and clays [73]. These sediments are layered on a metamorphic bedrock formed by impure fine to medium grain quartzites, black phyllites, slates, and schists (some graphitic) of the Paleozoic age, forming the lowest part of a mountain chain with elevations of approximately 2700 m a.s.l.

In the case of the Loja basin, the files of the Ecuadorian Secretary for Risk Management (SNGR) and the newspapers "La Hora" and "El Comercio" were reviewed to extract essential information on landslides caused by hydrological events. This review complemented previous research work [15] to extract essential information on landslides caused by hydrological events. Thereafter, it was possible to date up to 93 landslide events. The information reviewed did not contain specific data on the types and geometries of the catalogued landslides. However, from previous studies ([27] and references therein), it is known that the mass movements found in this area are commonly (85%) of the very slow or creep type, such as earth-slide or earth-flow, and the complex types of the Cruden and Varnes (1996) classification. They are very slow or creep landslides that accelerate gradually and become flows after high precipitation events. The analysis focused on these types of landslides because they are the most common and harmful in the study area. It also focused on applied superficial type landslides, whose fault planes reach up to approximately 30 m.

The most characteristic landslides in the Loja Valley have lengths ranging from 100 to 250 m and widths ranging from 60 to 150 m, which occur on slopes ranging from 10° to 40°. These are mainly associated with the clays, siltstones, and colluvial deposits of the Loja sedimentary basin (Figure 3). Mineralogical analyses show that the main factor causing landslides is the presence of active clay minerals in the geological formations involved [27].

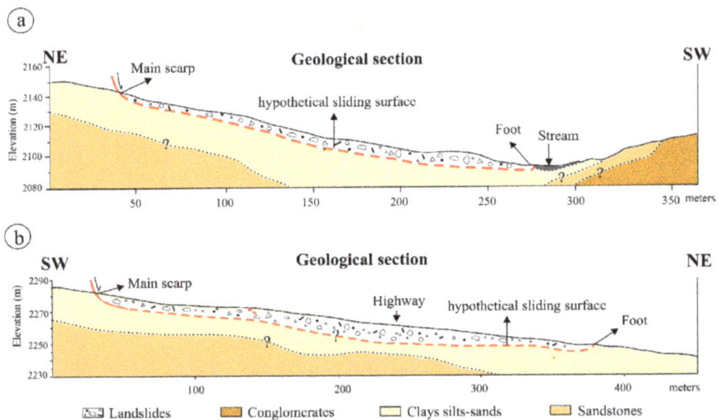

Figure 3. Examples of geological sections in the area of the Loja Basin, in the directions NE-SW (**a**) and in the direction SW-NE (**b**). Taken from [74].

3.2. Southern Granada (Southern Spain)

The area studied in the south of the province of Granada is located on the southern slopes of the Sierra Nevada in the Betic Cordillera with a general S–SW direction. The enclosed area, with an extension of about 2370 km^2, covers heights from 5 to 1450 m and extends between the meridians 3°07′ and 3°29′W and between the parallels 36°41′ and 36°′56N (Figure 1). The climate changes from arid-steppe to cold-arid (BSk) at the base of the mountains, to an arid-steppe hot-arid (BSh) zone on the coastal plain [75]. In this zone, the rainy and wet season is also related to the monthly interval running from October to April (Figure 2), with an average temperature of 16 °C and minimums at higher elevations. The summer period is dry, and temperatures can exceed 40 °C. The mean annual precipitation reaches 650 mm, and the main system controlling the weather is the Azores anticyclone. In winter, the Azores anticyclone high-pressure band is located between 40° and 30°N, leaving the progression of the Polar Front and Iceland Low moving downwards, which generates the majority of convections in this region [76]. However, in the warm-dry season, stormy weather is caused by the movement of the Azores anticyclone to the area between the latitudes 35° and 55°N, when the Polar Front and the Subtropical Jet cloud systems enter the study area. Elevation plays an important role in this zone as an altitudinal gradient that generates cold temperatures and a rainfall gradient [13]. The south of the province of Granada is located in the Internal Zone of the Betic Cordillera, which is mainly characterized by metamorphic rocks belonging to the Nevado-Filabride complex, composed of Triassic calc-schists, marbles, phyllites, and quartzites [77], and the Alpujarride complex, with dark schists and feldspathic mica-schists. These materials form the southern flank of the antiform of the Sierra Nevada, which has steep slopes. The lower part of these mountains is covered by post-tectonic Neogene and Quaternary deposits, such as marls, silts, and conglomerates. This lithology, combined with the study area geomorphometry of high slopes and deep valleys, is associated with different types of landslides [78,79]. The great number of mass movements in this area is represented by rapid synchronic landslides like rock falls and debris flow affecting the regolith in the upper soil layers and alluvial deposits downslope. However, a common case in the area consists of deeper seated quiescent (or dormant) landsides with intermittent or diachronic activity and inactive landslides. In addition, few cases are related to permanent slides with very slow displacements. Specifically, on the southern border of the part of the Sierra Nevada, which belongs to the Internal Zone, the highest level of instability has been observed, which is related to the periods with the highest precipitation rates. Two major rainfall periods leading to multiple landslide occurrences were seen in the wet seasons of 1996–1997 [80] and 2009–2010 [81] when large amounts of damage affected the roads and villages located in this area. The predominant types catalogued coincided with rockfalls and translational and rotational slides affecting jointed rock mass such as marble, phyllite, and schist, but also coincided with debris flows and complex movements, such as a combination of small translational slides evolving into debris flow. These complex movements and, in general, flow-like landslides developed on regolith and alluvial deposits.

Research has recently been carried out in both study areas with the aim of making progress and generating essential information to better understand the cause-effect relationship between the variance of rainfall and the activation of different types of landslides. The authors of [13,27] determined CRTs relating to mean intensity and duration extracted from rainfall events were causing landslides in the study area south of the province of Granada. In the case of Loja Basin, the relationship between duration and the accumulated rainfall for the rainfall events leading to landslides has been studied [27]. Using this previous work as a starting point, this paper seeks to compare both CRTs by adding the homologous relationship (accumulated rainfall vs. duration) to the case of the southern part of the province of Granada. This relationship has been used conventionally, although alternative CRT functions can be derived by applying simple transformations. For example, ID (Intensity-Duration) threshold curves can be obtained by applying the definition of the mean intensity (accumulated rainfall/duration). The second objective of this research is to correlate the rainfall of the humid season with the indices of climate cycles that create atypical rainfall conditions in these areas. To generate this information, first, the seven-month series are calculated.

Then, the time-series for the mean indices of the climate cycles are also calculated over the same range of humid months, so the positive and negative phases of rainfall and climate cycles can be compared. The graphical results are reported in this document and then contrasted in the discussion. This work will add important information to be utilized for hazard assessment and the predictive models of mass movements in a comparative way for two cases with different climates.

4. Materials and Methods

4.1. Materials

The baseline data are taken from the daily rainfall records available from previous work. These records are formed by using daily rainfall data (mm/24 h) gathered from:

- La Argelia rainfall gauge in the case of the Loja Basin, which presents the longest record for the area considered. These data were delivered by the INAMHI (Instituto Nacional de Meteorología e Hidrología). The time series collected runs from April 1965 to April 2015.
- Twelve rainfall gauges distributed throughout the study area in the south of the province of Granada. These data were provided by the Regional Water Agency Environmental Information Network and the National Meteorological Institute (AEMET). The times series collected have lengths starting from 1945, except for the cases of the meteorological stations S022, S220, S225, S392, and S447, which started in the years 1947, 1963, 1984, and 2000, respectively. Moreover, in general, these records reach the year 2011, except for the cases of S102, S153, and S220 ending in the years 2004, 2008, and 2010, respectively.

Six rainfall gauges are available in the Loja Basin (Figure 1); however, the only gauge providing a long and continuous record was that of the "La Argelia" meteorological station, so this was selected to extract the accumulated rainfall associated with each landslide event. The 12 rainfall gauges available in the area of southern Granada range between 300 and 1200 m a.s.l (Figure 4), except the rainfall gauges S154 and S225, which are located close to the coastal zone at altitudes of 33 and 60 m. Considering that the formation and extension of the precipitation cells can be affected by this altitude variability and the long dimension of the study area from west to east, the authors of [13] selected the closest rainfall gauges and used the longest records for each catalogued landslide to obtain the CRTs. However, this area constitutes a continuous and nearly homogeneous climate throughout its west-east dimension, composed of the southern face of the Sierra Nevada and the northern slope of the Guadalfeo River valley. This area is nearly homogeneous in its exposition to humid Atlantic winds [82]. It constitutes the windward of the humid winds leading to the known Föhn effect, when the water vapor is forced to rise by the orogenic obstacle through the windward side. Thus, the humid winds reach a sufficient height, resulting in adiabatic cooling and condensation that produce orographic precipitation, while in the northern part of the Sierra Nevada (leeward side) descending winds are drier and warmer. This effect is generalized throughout this study area so the mean monthly rainfall does not change significantly from one measuring site to another. By visually analyzing the mean monthly values for all the stations (Figure 4), it can be deduced that, except for the geographically lower gauges (S154 and S225), there are no significant differences.

However, to verify this effect, the ANOVA (Analysis of the Variance) was applied (Table 1) to contrast the variability of the monthly record registered on every rainfall gauge. By performing this analysis, the hypothesis of similar averages for the distribution of the different rainfall gauges can be accepted (p-value = 0.86). This fact was considered when analyzing the correlation between rainfall and climatic cycles, so averaged monthly rainfall values was obtained by taking into account all the rainfall gauges with the aim of using representative values for the entire area.

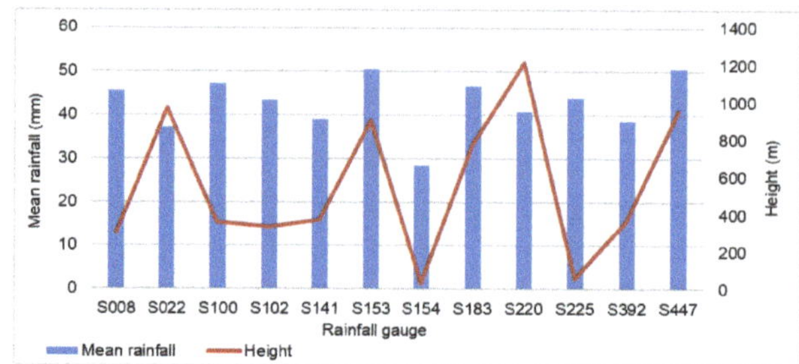

Figure 4. Mean values and heights of the rainfall gauges of southern Granada.

Table 1. Analysis of the variance (ANOVA) of the different rainfall gauges of southern Granada.

ANOVA						
Variance	**Square Sum (SS)**	**Degrees of Freedom (DF)**	**Square Mean (SM)**	**F**	**Probability**	**Critical Value of F**
Between groups	5209.23	11.00	473.57	0.56	0.86	1.86
Within groups	111,843.41	132.00	847.30			
Total	117,052.64	143.00				

The landslide catalogues were gathered as follows:

- In the case of the Loja Basin, the files from the Ecuadorian Secretary for Risk Management (SNGR) and "La Hora" and "El Comercio" newspapers were reviewed to extract essential information about landslides triggered by hydrological events. For the SNGR database, the data referring to the period from 2010 to 2015 were revised, while for the newspapers, the information available from 2002 was examined. After this review, up to 93 landslide events could be dated.
- In the case of southern Granada, the newspaper IDEAL provided the majority of the data related to landslides with its first issue from the 8 May 1932 [83]. This information was completed with the data recorded in earlier research work [80]. Given the inconveniences and limited information on landslides, only 20 landslides cases with the minimal information were found for this area.

In addition, the data series for climate indices were downloaded from the following sources:

- The monthly ENSO (El Niño Southern Oscillation) index for the period 1870–2019 from the NOAA.ESRL [84].
- The monthly NAO (North Atlantic Oscillation) index for the period 1950–2019 from NOAA.CPC [85].
- The monthly WeMO (Western Mediterranean Oscillation) index for the period 1821–2018 from the Climatological Group of the Barcelona University [86].

4.2. Methods

The procedure for analyzing the major differences between the two study areas regarding landslide triggering begins by calculating the CRT curves. The detailed methodology to extract the threshold values is presented in Palenzuela et al. [13], which permits the extraction of measurements on accumulated rainfall that coincides with extreme return periods (T). The start and end of the rainfall event is marked by the appearance of one or more consecutive T spikes. These spikes can coincide with a short period of intense rainfall, including the date of the landslide activation or, on the other hand, the anomalous T values can be associated with accumulated rainfall from the days before the date of the landslide. In the latter case, the lagged date of landslide activation is probably due to the time needed for the precipitation to infiltrate and saturate the soil layers above the slip plane. After determining the accumulated precipitation (A) as the magnitude of the rainfall event, its duration (D) and intensity (I) are calculated, as well as the associated return period (T) as the inverse of the frequency. The same

CRT connecting A and D is utilized in both study areas with the aim of contrasting the values of the hydrological variables and trends.

The second part of this analysis seeks to compare the relevant climate indices in each study area against the accumulated rainfall during the wettest seasons and the number of landslides collected. Considering the longer duration of the climate variability, which typically lasts about a year to several years [87,88], this interval has been selected to include the more humid period from October to April during which the greater number of landslides were also catalogued. Most of the landslide events collected are related to specific periods with anomalous rainfalls. In the Loja Basin, [27] identified a high number of landslides within the period 2014–2015 associated to 22 rainfall events with a greater return period (up to 3.5 years). Whereas, in the area of southern Granada, the landslides catalogued were concentrated between the wet seasons of 1996–1997 and 2009–2010 [13,80], some of them associated with accumulated rainfall values corresponding to a very low recurrence (up to 40 years). These considerations make it interesting to evaluate the degree of influence that the main climatic teleconnections affecting both study areas could have on the rainfall measured and, therefore, on the landsliding events. Accordingly, in this second part of the analysis, the accumulated rainfall of the wet seasons October to April for both study areas was assessed for the hydrological years of the entire time-series. These time-series were fitting them with linear functions representing central values that were subtracted to better show the variability (oscillation) of such variables. Thus, the greatest variations around 0 represent the positive (wettest periods) and negative spikes (drought periods). Regarding the climate indices, the mean values for the same period are calculated for the time-series of ENSO, NAO, and WeMO indices. After these time-series for the interval of October–April are prepared, their correlation with the accumulated rainfall is evaluated by using the Pearson coefficient.

5. Results

This section presents the main findings related to the characterization and contrasting of the rainfall events triggering landslides in both study areas, as well as the comparison between the variability of climate indices with the accumulated rainfall in the wettest rainy seasons. The outcomes are discussed in the Conclusion section.

5.1. Critical Rainfall Thresholds

The daily rainfall values corresponding to the landslide dates and the 89 days previous to them were automatically extracted from the rainfall series and added for durations ranging from 1 to 90 days by using VBA (Visual Basic) macros. This operation permitted the calculation of the A, D, I, and T measurements, as explained in Section 4.2, and in-depth in [13]. For comparison, the mean and standard deviation of the rainfall variables for both study areas are summarized in Table 2.

Table 2. Summary of the rainfall variables for the events triggering catalogued landslides. Aver. and SD stand for average and standard deviation, respectively.

Geolocation	Climatic Zone	Rainfall (E)		Duration (E)	
		Aver.	SD	Aver.	SD
Loja (southern Ecuador)	Cfb	156.4 mm	115.2 mm	25.3 d	28.93 d
Granada (southern Spain)	BSk	458.1 mm	233.5 mm	42.8 d	22.39 d

The resulting values of accumulated rainfall and duration for each event generating landslides were represented in a scatter plot, a power-law curve was fitted (Table 3), and the confidence and prediction bounds were added to represent the CRTs for every case. The minimum values of the scatter plot were manually selected to adjust the threshold curve in every case. In addition, the statistical significance was determined by obtaining the determination coefficient R^2. In the Loja Basin, a generalization was given for the landslide type as that of very slow mass movements that increased their displacement rates as a consequence of the rainfall events.

Table 3. Summary of CRT (Critical Rainfall Thresholds) parameters and statistical significance of the curve fit.

Area	Type	α	β	R^2
Loja	Earth-slide, earth-flow, complex	7.33	0.76	1.02
Granada	Translational slide	92.35	0.15	0.33
Granada	Rock fall	63.74	0.39	1.00
Granada	Complex	52.34	0.42	1.33
Granada	All types	64.36	0.22	0.93

This led to a greater amount of data (Figure 5a) to apply to the curve-fitting process, which is shown by its determination coefficient $R^2 = 1$. In the case of southern Granada, the information revised permitted the mass movements to be divided into three types: translational slides (5) (Figure 5b), rock falls (6) (Figure 5c), and complex landslides (8) (Figure 5d), although the mixed case was also considered (Figure 5e). In this case, the CRT function with the lowest statistical significance corresponded to the translational slides, as a very low number of cases were collected for this type with a high dispersion. On the contrary, rock falls and complex mass movements, as well as the mixed case, show high statistical significance. By comparing both study areas, the scaling factor α is lowest in the case of the Loja Basin while the shape factor β coincides with the highest value. This means that the lowest values are needed to activate or reactivate mass movements in this area, but with the increase in duration, the cumulative rainfall increases more quickly than in the case of southern Granada. The different landslide types of the southern Granada area show that translational slides are activated by the highest cumulative rainfall values, while and rock falls and complex slides shows similar activation patterns. As expected, the case involving all the mass movements in southern Granada results in a conservative curve with relative low values when compared with the CRT curves of separated types.

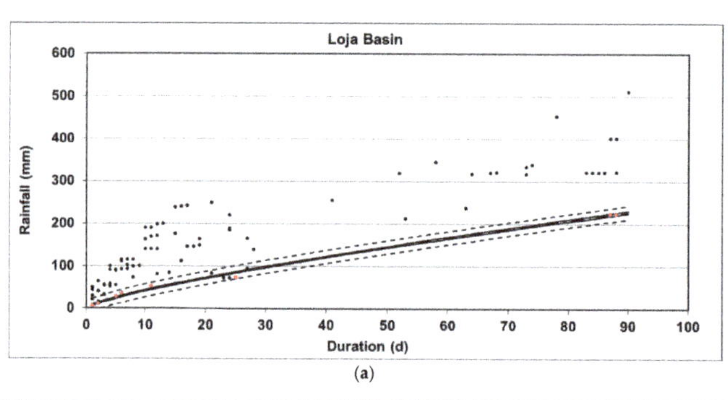

(a)

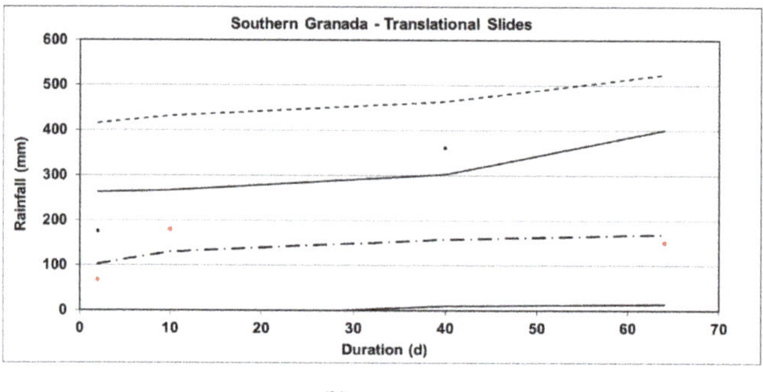

(b)

Figure 5. *Cont.*

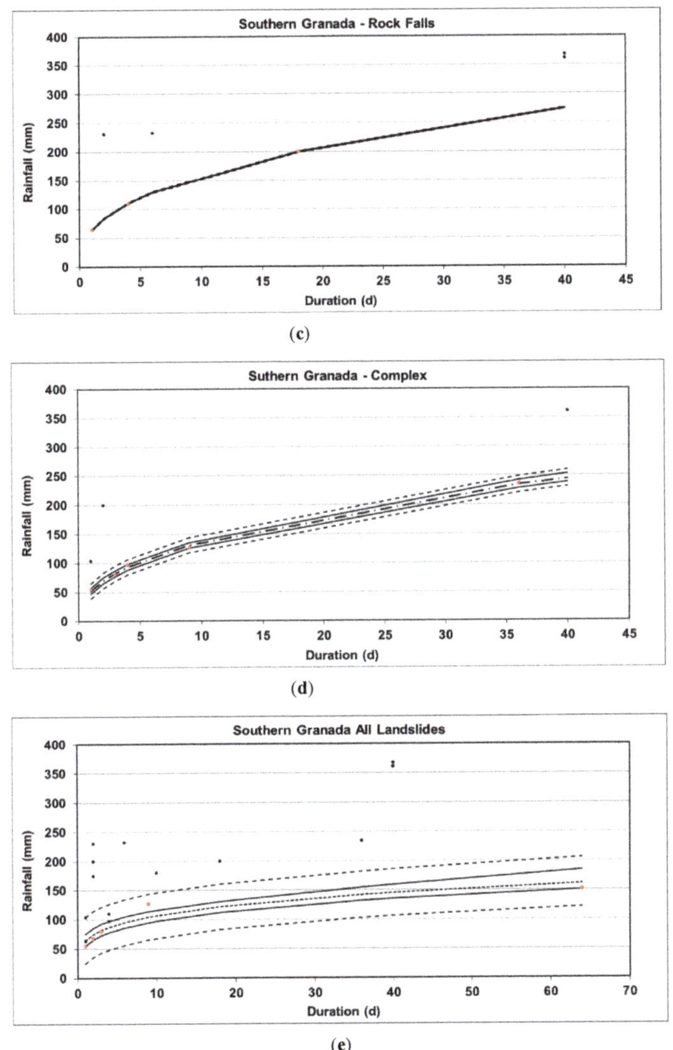

Figure 5. Scatter plots of the rainfall thresholds and the fitted CRTs: (**a**) fitted curve for low to very low mass movements in the Loja Basin (modified from [5]); (**b–e**) are fitted curves for the translational slides, rock falls, complex landslides, and all the landslide types, respectively, in southern Granada. Dash-dotted line: curve fit to lower values Rainfall-Duration; solid line: confidence bounds; dashed line: prediction bounds. Red points are the manually selected points to fit the CRT curve.

5.2. Correlation Between Teleconnections and Accumulated Rainfall

The accumulated rainfall observed for the wet season October–April was extracted from the time series and plotted together with the more significant short frequency climate indices in the study areas. The visual information shows that, in general, the correlation between the wet seasonal rainfall with the oscillation shown by the climatic-cycle indices is low (from 0.27 to 0.55, as absolute values). The rain gauge of the La Argelia meteorological station was used for the Loja Basin, while for southern Granada, the global average of all the rainfall gauges spatially distributed in the study area was calculated. In the case of the Loja Basin, the positive and negative phases of the accumulated rainfall show numerous coincidences (in phase peaks) with the ENSO index (Figure 6a), and this is confirmed by the positive correlation of 0.27 (Table 2). However, this pattern is not so clear at a monthly scale (Figure 6b). A strong matching can be observed in the periods 1982–1984 and 1999–2000 when the highest rainfall peaks were detected. With regard to the number of landslides, the highest spikes of this variable do not show a clear matching with either the peaks of the ENSO index or the accumulated rainfall. For example, the number of landslides (standardized from 0 to 4 in Figure 6a,b) registered from 2005 to 2010 are equally distributed, independent of the positive and negative peaks of the ENSO index, whereas the peak of the wet season 2007–2008 coincides with a relative maximum (high peak) of the climatic index. On the contrary, the high number of landslides of December 2012 is correlated with a relative minimum of the ENSO. Similarly, two peaks of the relative number of landslides catalogued in December 2012 and December 2014 coincide with two maximums of accumulated rainfall in the wet season. However, the high rainfall peak of 2007–2008, coinciding with a positive peak of the ENSO index, shows a low landslide count. In the case of southern Granada, the visual analysis shows many of the relative minimums and maximums of NAO coinciding with the opposite phases of the variation of rainfall (e.g., in yearly intervals from 1962–1964, 1968–1971, 1971–1973, 1991–1995, 1995–1997, and 2009–2011) (Figure 6c), but there are also numerous peaks with the same phase, which explains the low correlation coefficient (Table 4). The same relationship, although lower, is shown by the graphs with the WeMO (Figure 6d). However, when the two phases of NAO and WeMO are added together, the correlation with the accumulated rainfall reaches its highest value (Figure 6e), with a Pearson coefficient of −0.55.

Table 5 shows specific values for the hydrological years with well-documented landslides in the zone of Granada. In this case, the two periods with the maximum number of landslides generated during extremely humid seasons (1996–1997 and 2008–2009) fall within the amplitude of two significant minimums (negative spikes) of the NAO and NAO + WeMO indices.

Table 4. Pearson coefficients for the correlation between the different climate indices and the differences after subtracting the general trend from the observed accumulated rainfall for the wet seasons October–April.

Geolocation	Climate Index	Pearson Coef.
Loja (southern Ecuador)	ENSO	0.27
Granada (southern Spain)	NAO	−0.44
	WeMO	−0.31
	NAO + WeMO	−0.55

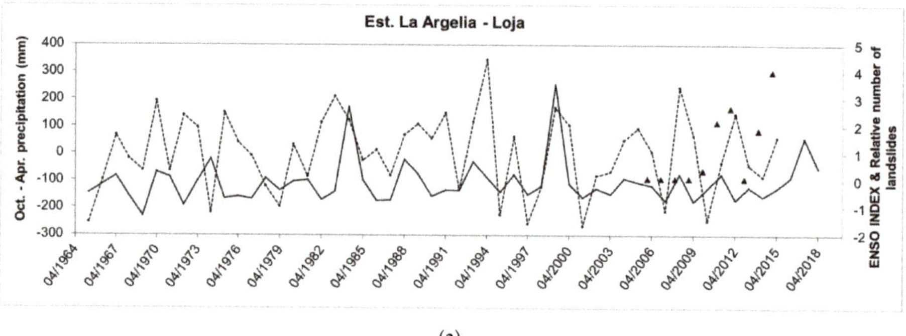

(a)

Figure 6. *Cont.*

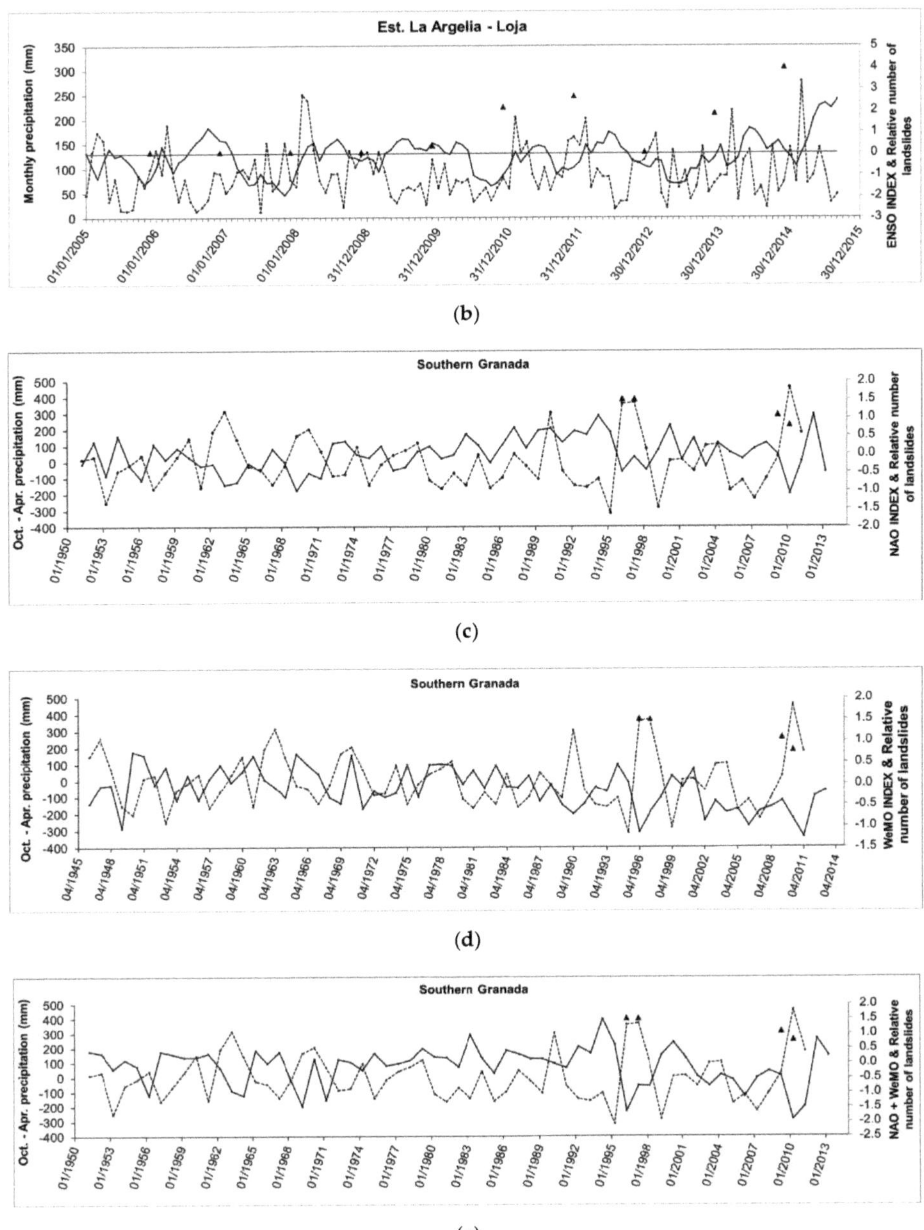

Figure 6. Graph of the differences after subtracting the general trend from the accumulated rainfall observed for the wet seasons October–April (dotted line) and the more significant climate indices (solid line): (**a**) graph for the Loja Basin including the ENSO index; (**b**) graph for the Loja Basin including the ENSO index for the monthly scale and the regression line of the ENSO index; (**c**) graph southern Granada including the NAO index; (**d**) graph for southern Granada including the WeMO index; (**e**) graph for southern Granada including the aggregation of NAO and WeMO indices. Triangles represent the relative quantity of documented landslides by year.

Thus, in general, low correlations were identified and a well-defined cause-effect link cannot be determined. This fact can be attributed to one or both reasons: a) the limitation of the existing data, such as incomplete landslide datasets, and b) the complex and changing patterns of the climatic teleconnections.

Table 5. Magnitude of the negative spikes for the climate indices in southern Granada in hydrological years with documented landslides.

Hydrological Year	Index	Values	% Respect to Average of Negative Values
	NAO	−0.49	131.17
1995–1996	WeMO	−1.15	266.82
	NAO + WeMO	−1.65	254.9
	NAO	−0.09	24.64
1996–1997	WeMO	−0.69	159.56
	NAO + WeMO	−0.78	121.04
	NAO	−0.03	8.34
2008–2009	WeMO	−0.42	96.33
	NAO + WeMO	−0.45	69.26
	NAO	−1.10	290.76
2009–2010	WeMO	−0.84	193.66
	NAO + WeMO	−1.93	299.16
Average:	NAO	−0.38	
	WeMO	−0.43	
	NAO + WeMO	−0.65	
1st quartile	NAO	−0.50	
	WeMO	−0.61	
	NAO + WeMO	−1.07	

6. Conclusions

Despite the difficulties hampering progress and accurate results, this work is in line with the main practices in landslide risk mitigation and reduction and provides a contribution to the knowledge generation on landslide hazards. Specifically, this paper analyzes the relationship between landslide development and the climate variability in two mountainous sites of southern Ecuador and southern Spain. As a result, the information derived from this experimental research serves as a reference base in the prevention of hydrological and climatological conditions that most likely lead to landslides of different typology. In addition, a result comparison is depicted from the two areas with different climate and geological settings.

In view of the comparative findings, it is clear that in the case of the humid climate of the Loja Basin with thick layers of residual soils and meteorized rock, landslides occur with lower cumulative rainfall. The average magnitude for the Loja case is in the order of 34% of the mean cumulative rainfall in southern Granada. Similarly, the mean duration of the rainfall events that include the antecedent rainfall until the date on which landslides are reported is slightly higher for Loja than the mean duration in the case of southern Granada. This causes the higher recurrence of landslide events in the Loja Basin, as reported in Soto et al. [27], and only 24% of the rainfall return periods exceeded 1 year. On the contrary, Palenzuela et al. [13] stated that the return period of the first quartile of the catalogued landslides did not exceeded 1.1 years. The extraction of the major rainfall characteristics (accumulated rainfall and duration) related to the activation of landslides was used to estimate the CRTs for both cases. After fitting the power law functions, it was revealed that for the Loja Basin, which has a predominance of very low mass movements, the lowest minimum values are expected to activate/reactivate these landslides when compared with those of southern Granada. This finding is probably related to the wetter climate and greater number of rainy days, which favors a high saturation degree acting as a predisposing factor. In the case of southern Granada, although a lower number of landslides was catalogued, two of the types analyzed—rock fall and complex mass movements—showed maximum statistical significance, while translational slides were weakly fitted due to the minimum number of catalogued events and their plotting dispersion. Under this relationship, the greater activation values of cumulative rainfall were obtained for the case of translational slides with a smooth trend corresponding with a low β value ($\beta = 0.15$). On the contrary, lower thresholds are attributed to rock fall, complex landslides, and the global case, including every type of mass movement. This is probably due to the shallower slip-planes and the low shear strength of the materials involved in rock falls and complex slides (mainly shallow slides evolving to debris flow). The curve parametrization also shows similar values in the cases of rock fall and complex landslides, although with higher α for the first type. In addition, the accumulated rainfall for the wettest period for both study cases (October–April) was calculated for all the hydrological years of the precipitation series. The general trend was then subtracted to the new time series from both study areas to highlight their variability. Then, the major climate indices were correlated with the remaining component. Although shorter or longer intervals could be tried, this interval has been selected with the aim of getting insights of the correlation within the more humid period from October to April, coinciding with the period when most of the landslides were catalogued. The findings show a generalized low correlation, although with some matching between the phases of the climatic oscillations and the accumulated rainfall, as well as with the number of landslides. After analyzing the results for the time series of the Loja Basin, low correlation was revealed. The ENSO index presents numerous hydrological years with in-phase coincidences between accumulated rainfall and the ENSO index, but a significant number of cases do not follow this pattern, which makes the degree of correlation decrease. More uncommonly, there are observations between high rainfall peaks and the number of landslides that show a lack of events, or a nearly constant number of them in years when the accumulated rainfall and/or the ENSO index are higher. For the series of southern Granada, the independent correlations between te accumulated rainfall and the indices of NAO and WeMO show negative coefficients. However, the strength of the correlation increases with the combined NAO and WeMO indices, giving a Pearson coefficient of −0.55. This preliminary information is important for hydrological risk management; however, there is low correlation between large scale teleconnections and extreme rainfall that triggered landslides. The low correlation between both variables is probably linked to the complex and non-persistent spatial-temporal patterns of all of the climatic phenomena [29], which makes difficult to understand their history and the causes for their recurrence and intensity. Thus, as stated in the literature [89–91] for the ENSO phenomenon, this fact can generate differences in the phases and intensities at the global and local scales. Moreover, terrain features make these patterns even more complex and less uniform in mountain ranges as rainfall cells strongly depend largely on altitude, topographic barriers, and slope expositions [40]. Similarly, the lack of landsliding events in the wet seasons when extreme rainfall was determined is probably due on the one hand to the information biased in space and time. On the other hand, precipitation series cover the dates of the landslide occurrences and antecedent periods and they are considerably long for calculating rainfall frequencies and return periods. However, a longer sampled period for landsliding events would be more suitable for the application of the statistical approach to evaluate this correlation [31], providing more responses of positive (associated with landslides) and negative rainfall events. Thus, the implementation of appropriate practices on landslide recording would improve drastically this type the reliability and accuracy of this type of correlations. Additionally, the measurement of the horary rainfall depth and the monitoring of the piezometric level are of significant importance as a part of alert systems or prediction models at the scale of study-site cases.

In conclusion, a minimal correlation can be determined between climatic phenomena affecting a significant part of global geography and the extreme rainfall-triggered landslides, but results have to be considered as preliminary insights as more detailed information is necessary to accurately define these patterns. The major inconvenience found in this research has been related to the scarce data on landsliding that can be indirectly obtained. In the case of the Loja Basin, the number of landslides dated was higher, but no detailed information could be gathered on the type, thickness, or volume and other characteristics. On the contrary, in the case of southern Granada, the number of landslides collected was lower, but some classes could be assigned to them. Thus, for the future advancements in landsliding hazard and prediction, systematic and appropriate practices to monitor both landslides and rainfall in the spatial and time dimensions is highly recommended in the study areas. This can be developed by implementing recently developed techniques like change detection analysis, or by exploiting the high number of daily satellite images that are currently being recorded and applying automatic and semi-automatic techniques. Nowadays, the integration of mobile and geospatial technologies and paradigms such as Citizen Science (CitSci) and VGI (Volunteered Geographical Information) with the automatic image processing through the implementation of AI (Artificial Intelligence) based models appears as a power instrument to collect new environmental data [92]. Admittedly, to gain statistical significance, an efficient plan for cataloguing and registering in situ data on the type, dimensions, and geotechnical parameters of envisaged landslides is needed, as it is of major importance to calibrate rainfall threshold curves. In addition, the application of time-series decomposition into trends and periodicities (or oscillations) is recommendable to make progress in the prediction of extreme rainfall recurrence. Currently, the modelling and prediction of time-series has been approached by using new advanced data-based models such as soft-computing models or chaotic based models providing the advantage for overcoming the problems of non-linear and non-stationary time-series [93,94]. Additionally, improvements on the accuracy of the prediction on rainfall patterns can be addressed by applying novel precipitation models. However, where landslides have a high recurrence, like in the case of the Loja Basin, forecast models integrating landslide thresholds with physical susceptibility models (i.e., the Hydrological-Geotechnical model) will have an even more important role in the mitigation of different type of landslides.

Author Contributions: Data curation, J.S.L.; Investigation, J.A.P.B.; Methodology, J.A.P.B.; Writing—original draft, J.A.P.B.; Writing—review & editing, J.S.L. and C.I.F. All authors have read and agreed to the published version of the manuscript.

Funding: This research received no external funding

Acknowledgments: This research was supported by the project CGL2008-04854 funded by the Ministry of Science and Education of Spain and by the Ministry of Higher Education, Science, Technology and Innovation (SENESCYT) under the scholarship program "Open Call 2012 Second Phase" of the government of Ecuador. Authors want to acknowledge the Ecuadorian National Meteorological and Hydrologic Institute (INAMHI) and the Ecuadorian Secretary for Risk Management (SNGR-Zone 7) for providing the rainfall records and part of the landslide inventory, respectively.

Conflicts of Interest: The authors declare no conflict of interest.

References

1. UNESCO. *Annual Summaries of Information on Natural Disasters, 1971–1975*; UNESCO: Paris, France, 1979.
2. Rib, H.T.; Liang, T. Recognition and identification. In *Landslides Analysis and Control*; Schuster, R.L., Krizek, R.J., Eds.; Washington Transportation Research Board, Special Report; National Academy of Sciences: Washington, DC, USA, 1978; Volume 176, pp. 34–80.
3. Ayala, F.J.; Elizaga, E.; de González Vallejo, L.I. *Impacto Económico y Social de los Riesgos Geológicos en España*; ITGE: Madrid, Spain, 1987; p. 134.
4. Suárez, R.R.; Regueiro, M. *Guía Ciudadana de los Riesgos Geológicos*; I.C.O.G: Madrid, Spain, 1997.
5. Petley, D. Global patterns of loss of life from landslides. *Geology* **2012**, *40*, 927–930. [CrossRef]
6. Spizzichino, D.; Margottini, C.; Trigila, A.; Iadanza, C. Landslide impacts in europe: Weaknesses and strengths of databases available at european and national scale. In *Landslide Science and Practice*; WLF 2011, Rome; Springer: Berlin/Heidelberg, Germany, 2013; pp. 73–80.
7. White, I.D.; Mottershead, D.N.; Harrison, J.J. *Environmental Systems*, 2nd ed.; Chapman & Hall: London, UK, 1996; p. 616.
8. Aleotti, P. A warning system for rainfall-induced shallow failures. *Eng. Geol.* **2004**, *73*, 247–265. [CrossRef]
9. Wieczorek, G.; Glade, T. Climatic factors influencing occurrence of debris flows. In *Debris-Flow Hazards and Related Phenomena*; Springer: Berlin/Heidelberg, Germany, 2005; pp. 325–362. [CrossRef]
10. Guzzetti, F.; Peruccacci, S.; Rossi, M.; Stark, C.P. Rainfall thresholds for the initiation of landslides in central and southern Europe. *Meteorol. Atmos. Phys.* **2007**, *98*, 239–267. [CrossRef]

11. Guzzetti, F.; Peruccacci, S.; Rossi, M.; Stark, C. The rainfall intensity–duration control of shallow landslides and debris flows: An update. *Landslides* **2008**, *5*, 3–17. [CrossRef]
12. Vennari, C.; Gariano, S.L.; Antronico, L.; Brunetti, M.T.; Iovine, G.; Peruccacci, S.; Terranova, O.; Guzzetti, F. Rainfall thresholds for shallow landslide occurrence in Calabria, southern Italy. *Nat. Hazards Earth Syst. Sci.* **2014**, *14*, 317–330. [CrossRef]
13. Palenzuela, J.A.; Jiménez-Perálvarez, J.D.; Chacón, J.; Irigaray, C. Assessing critical rainfall thresholds for landslide triggering by generating additional information from a reduced database: An approach with examples from the Betic Cordillera (Spain). *Nat. Hazards* **2016**, *84*, 185–212. [CrossRef]
14. Ibsen, M.-L.; Brunsden, D. The nature, use and problems of historical archives for the temporal occurrence of landslides, with specific reference to the south coast of Britain, Ventnor, Isle of Wight. *Geomorphology* **1996**, *15*, 241–258. [CrossRef]
15. Caine, N. The Rainfall Intensity: Duration Control of Shallow Landslides and Debris Flows. *Geogr. Ann. Ser. APhys. Geogr.* **1980**, *62*, 23–27. [CrossRef]
16. Innes, J.L. Debris flows. *Prog. Phys. Geog.* **1983**, *7*, 169–501. [CrossRef]
17. Corominas, J.; Moya, J. Reconstructing recent landslide activity in relation to rainfall in the Llobregat River basin, Eastern Pyrenees, Spain. *Geomorphology* **1999**, *30*, 79–93. [CrossRef]
18. Zêzere, J.; Rodrigues, M. Rainfall Thresholds for Landsliding in Lisbon Area (Portugal). In Proceedings of the Conference: 1st European Conference on Landslides, Lisse, Prague, 24–26 June 2002.
19. Annunziati, A.; Focardi, A.; Focardi, P.; Martello, S.; Vannocci, P. Analysis of the rainfall thresholds that induced debris flows in the area of Apuan Alps Tuscany, Italy (19 June 1996 storm). In Proceedings of the EGS Plinius Conference on Mediterranean Storms, Maratea, Italy, 14–16 October 1999.
20. Giannecchini, R. Rainfall triggering soil slips in the southern Apuan Alps (Tuscany, Italy). *Adv. Geosci.* **2005**, 2. [CrossRef]
21. Govi, M.; Mortara, G.; Sorzana, P. Eventi idrologici e frane. *Geol. Appl. Ing.* **1985**, *20*, 359–375.
22. Cardinali, M.; Galli, M.; Guzzetti, F.; Ardizzone, F.; Reichenbach, P.; Bartoccini, P. Rainfall induced landslides in December 2004 in south-western Umbria, central Italy: Types, extent, damage and risk assessment. *Nat. Hazards Earth Syst. Sci.* **2006**, *6*, 237–260. [CrossRef]
23. Terlien, M.T.J. The determination of statistical and deterministic hydrological landslide-triggering thresholds. *Environ. Geol.* **1998**, *35*, 124–130. [CrossRef]
24. Pasuto, A.; Silvano, S. Rainfall as a triggering factor of shallow mass movements. A case study in the Dolomites, Italy. *Environ. Geol.* **1998**, *35*, 184–189. [CrossRef]
25. De Vita, P. Fenomeni di instabilità delle coperture piroclastiche dei Monti Lattari, di Sarno e di Salerno (Campania) ed analisi degli eventi pluviometrici determinanti. *Quad. Geol. Appl.* **2000**, *7*, 213–235.
26. Chleborad, A.F. *Preliminary Evaluation of a Precipitation Threshold for Anticipating the Occurrence of Landslides in the Seattle, Washington, Area*; Open-File Report 03-463; US Geological Survey: Reston, VA, USA, 2003.
27. Soto, J.; Palenzuela, J.A.; Galve, J.; Luque-Espinar, J.A.; Azañón, J.; Tamay, J.; Irigaray, C. Estimation of empirical rainfall thresholds for landslide triggering using partial duration series and their relation with climatic cycles. An application in southern Ecuador. *Bull. Eng. Geol. Environ.* **2017**. [CrossRef]
28. Walker, G.T. Correlation in seasonal variation of weather. VIII: A preliminary study of world weather. *Mem. India Meteorol. Dep.* **1923**, *24*, 75–131.
29. HF, D.; GN, K. Atmosferic teleconnections associated with the extremes phases of Southern Oscillation. In *El Niño Historical and Paleoclimatic Aspects of the Southern Oscillation*; Diaz, M., Ed.; Cambridge University Press: Cambridge, UK, 1992; pp. 7–28.
30. Salinger, M.J.; Lefale, P. The occurrence and predictability of extreme events over the southwest pacific with particular reference to ENSO. In *Natural Disasters and Extreme Events in Agriculture: Impacts and Mitigation*; Springer-Verlag: Heidelberg/Berlin, Germany, 2005; pp. 34–49. [CrossRef]
31. Froude, M.J.; Petley, D. Global fatal landslide occurrence from 2004 to 2016. *Nat. Hazards Earth Syst. Sci.* **2018**, *18*, 2161–2181. [CrossRef]
32. Sepúlveda, S.A.; Petley, D.N. Regional trends and controlling factors of fatal landslides in Latin America and the Caribbean. *Nat. Hazards Earth Syst. Sci.* **2015**, *15*, 1821–1833. [CrossRef]
33. Grimm, A.M.; Tedeschi, R.G. ENSO and Extreme Rainfall Events in South America. *J. Clim.* **2009**, *22*, 1589–1609. [CrossRef]

34. Aparicio-Effen, M.; Arana, I.; Aparicio, J.; Cortez, P.; Coronel, G.; Pastén, M.; Nagy, G.J.; Galeano Rojas, A.; Flores, L.; Bidegain, M. Introducing Hydro-Climatic Extremes and Human Impacts in Bolivia, Paraguay and Uruguay. In *Climate Change and Health: Improving Resilience and Reducing Risks*; Leal Filho, W., Azeiteiro, U.M., Alves, F., Eds.; Springer International Publishing: Cham, Switzerland, 2016; pp. 449–473. [CrossRef]
35. Hoyos, I.; Dominguez, F.; Cañón-Barriga, J.; Martínez, J.A.; Nieto, R.; Gimeno, L.; Dirmeyer, P.A. Moisture origin and transport processes in Colombia, northern South America. *Clim. Dyn.* **2018**, *50*, 971–990. [CrossRef]
36. Aristizábal, E. Influencia del ENSO en la variabilidad espacial y temporal de la ocurrencia de movimientos en masa detonados por lluvias en la región Andina. *Ingeniería y Ciencia* **2019**. [CrossRef]
37. Vergara Dal Pont, I.P.; Santibañez Ossa, F.A.; Araneo, D.; Ferrando Acuña, F.J.; Moreiras, S.M. Determination of probabilities for the generation of high-discharge flows in the middle basin of Elqui River, Chile. *Nat. Hazards* **2018**, *93*, 531–546. [CrossRef]
38. Moreiras, S.M. Climatic effect of ENSO associated with landslide occurrence in the Central Andes, Mendoza Province, Argentina. *Landslides* **2005**, *2*, 53–59. [CrossRef]
39. Hermanns, R.L.; Niedermann, S.; Ivy-Ochs, S.; Kubik, P.W. Rock avalanching into a landslide-dammed lake causing multiple dam failure in Las Conchas valley (NW Argentina)—Evidence from surface exposure dating and stratigraphic analyses. *Landslides* **2004**, *1*, 113–122. [CrossRef]
40. Moreiras, S.M.; Pont, I.V.D.; Araneo, D. Were merely storm-landslides driven by the 2015-2016 Niño in the Mendoza River valley? *Landslides* **2018**, *15*, 997–1014. [CrossRef]
41. Garcia-Herrera, R.; Barriopedro, D.; Hernández, E.; Diaz, H.F.; Garcia, R.R.; Prieto, M.R.; Moyano, R. A Chronology of El Niño Events from Primary Documentary Sources in Northern Peru. *J. Clim.* **2008**, *21*, 1948–1962. [CrossRef]
42. Vilímek, V.; Hanzlík, J.; Sládek, I.; Šandov, M.; Santillán, N. The Share of Landslides in the Occurrence of Natural Hazards and the Significance of El Niño in the Cordillera Blanca and Cordillera Negra Mountains, Peru. In *Landslides: Global Risk Preparedness*; Sassa, K., Rouhban, B., Briceño, S., McSaveney, M., He, B., Eds.; Springer: Berlin/Heidelberg, Germany, 2013; pp. 133–148. [CrossRef]
43. Rossel, F.; Mejia, R.; Ontaneda, G.; Pombosa, R.; Roura, J.; Le Goulven, P.; Cadier, E.; Calvez, R. Regionalization of El Niño influence on rainfall in Ecuador. *Bulletin de L'Institut Francaise d'Etudes Andines* **1998**, *27*, 643–654.
44. Lagos, P.; Silva, Y.; Nickl, E. El Niño y la Precipitación en los Andes del Perú. *Sociedad Geológica del Perú Lima Volumen jubilar en honor a Alberto Giesecke Matto* **2005**, *6*, 7–23.
45. Son, R.; Wang, S.Y.S.; Tseng, W.-L.; Barreto Schuler, C.W.; Becker, E.; Yoon, J.-H. Climate diagnostics of the extreme floods in Peru during early 2017. *Clim. Dyn.* **2020**, *54*, 935–945. [CrossRef]
46. De Guenni, L.B.; García, M.; Muñoz, Á.G.; Santos, J.L.; Cedeño, A.; Perugachi, C.; Castillo, J. Predicting monthly precipitation along coastal Ecuador: ENSO and transfer function models. *Theor. Appl. Climatol.* **2017**, *129*, 1059–1073. [CrossRef]
47. Trigo, R.M.; Osborn, T.J.; Corte-Real, J.M. The North Atlantic Oscillation influence on Europe: Climate impacts and associated physical mechanisms. *Clim. Res.* **2002**, *20*, 9–17. [CrossRef]
48. Trigo, R.M.; Pozo-Vázquez, D.; Osborn, T.J.; Castro-Díez, Y.; Gámiz-Fortis, S.; Esteban-Parra, M.J. North Atlantic oscillation influence on precipitation, river flow and water resources in the Iberian Peninsula. *Int. J. Climatol. A J. R. Meteorol. Soc.* **2004**, *24*, 925–944. [CrossRef]
49. Vicente-Serrano, S.M.; López-Moreno, J.I. Nonstationary influence of the North Atlantic Oscillation on European precipitation. *J. Geophys. Res. Atmos.* **2008**, 113. [CrossRef]
50. Hurrell, J.W. Decadal Trends in the North Atlantic Oscillation: Regional Temperatures and Precipitation. *Science* **1995**, *269*, 676–679. [CrossRef] [PubMed]
51. Hurrell, J.W.; Deser, C. North Atlantic climate variability: The role of the North Atlantic Oscillation. *J. Mar. Syst.* **2010**, *79*, 231–244. [CrossRef]
52. Jones, P.D.; Jonsson, T.; Wheeler, D. Extension to the North Atlantic oscillation using early instrumental pressure observations from Gibraltar and south-west Iceland. *Int. J. Climatol. A J. R. Meteorol. Soc.* **1997**, *17*, 1433–1450. [CrossRef]
53. Santos, M.; Fonseca, A.; Fragoso, M.; Santos, J.A. Recent and future changes of precipitation extremes in mainland Portugal. *Theor. Appl. Climatol.* **2019**, *137*, 1305–1319. [CrossRef]
54. Andrade, C.; Leite, S.; Santos, J. Temperature extremes in Europe: Overview of their driving atmospheric patterns. *Nat. Hazards Earth Syst. Sci.* **2012**, *12*, 1671. [CrossRef]

55. Santos, J.; Corte-Real, J. Temperature extremes in Europe and wintertime large-scale atmospheric circulation: HadCM3 future scenarios. *Clim. Res.* **2006**, *31*, 3–18. [CrossRef]
56. Santos, J.A.; Corte-Real, J.; Ulbrich, U.; Palutikof, J. European winter precipitation extremes and large-scale circulation: A coupled model and its scenarios. *Theor. Appl. Climatol.* **2007**, *87*, 85–102. [CrossRef]
57. Santos, M.; Fragoso, M.; Santos, J.A. Damaging flood severity assessment in Northern Portugal over more than 150 years (1865–2016). *Nat. Hazards* **2018**, *91*, 983–1002. [CrossRef]
58. Benito, G.; Brázdil, R.; Herget, J.; Machado, M.J.; Discussions, E.S.S. Quantitative historical hydrology in Europe. *Hydrol. Earth Syst. Sci.* **2015**, *19*, 3517–3539. [CrossRef]
59. Kiss, A. Introduction: Floods and Water-Level Fluctuations in Medieval (Central-)Europe. In *Floods and Long-Term Water-Level Changes in Medieval Hungary*; Kiss, A., Ed.; Springer International Publishing: Cham, Switzerland, 2019; pp. 1–50. [CrossRef]
60. De Lima, M.I.P.; Santo, F.E.; Ramos, A.M.; Trigo, R.M. Trends and correlations in annual extreme precipitation indices for mainland Portugal, 1941–2007. *Theor. Appl. Climatol.* **2015**, *119*, 55–75. [CrossRef]
61. Mariotti, A.; Zeng, N.; Lau, K.-M. Euro-Mediterranean rainfall and ENSO—A seasonally varying relationship. *Geophys. Res. Lett.* **2002**, *29*, 59-1–59-4. [CrossRef]
62. Lionello, P.; Malanotte-Rizzoli, P.; Boscolo, R. *Mediterranean Climate Variability*; Elsevier: Amsterdam, The Netherlands, 2006.
63. Brönnimann, S.; Xoplaki, E.; Casty, C.; Pauling, A.; Luterbacher, J.J.C.D. ENSO influence on Europe during the last centuries. *Clim. Dyn.* **2007**, *28*, 181–197. [CrossRef]
64. Chacón, J.; Alameda-Hernández, P.; Chacón, E.; Delgado, J.; El Hamdouni, R.; Fernández, P.; Fernández, T.; Gómez-López, J.M.; Irigaray, C.; Jiménez-Perálvarez, J.; et al. The Calaiza landslide on the coast of Granada (Andalusia, Spain). *Bull. Eng. Geol. Environ.* **2019**, *78*, 2107–2124. [CrossRef]
65. Luque-Espinar, J.A.; Mateos, R.M.; García-Moreno, I.; Pardo-Igúzquiza, E.; Herrera, G. Spectral analysis of climate cycles to predict rainfall induced landslides in the western Mediterranean (Majorca, Spain). *Nat. Hazards* **2017**, *89*, 985–1007. [CrossRef]
66. Mateos, R.; García-Moreno, I.; Azañón, J.; Tsige, M. La avalancha de rocas de Son Cocó (Alaró, Mallorca). Descripción y análisis del movimiento. *Boletín Geológico y Minero* **2010**, *121*, 153–168.
67. Martín-Vide, J. Ensayo sobre la Oscilación del Mediterráneo Occidental y su influencia en la pluviometría del este de España. In Proceedings of the III Congreso de la Asociación Española de Climatología "El agua y el clima", Palma de Mallorca, Spain, 16–19 June 2002.
68. Lemus Cánovas, M.; López-Bustins, J.A. Variabilidad espacio-temporal Inde la precipitación en el sur de Cataluña y su relación con la oscilación del Mediterráneo Occidental (WeMO). In Proceedings of the X International Congress AEC: Clima, sociedad, riesgos y ordenación del territorio, Alicante, Spain, 5–8 October 2016; Volume 21, pp. 225–236.
69. Köeppen. Das geographische system der klimate. In *Handbuch der Klimatologie, 1C.*; Köeppen, W., Geiger, R., Eds.; Gebrüder Borntraeger: Berlin, Germany, 1936; p. 44.
70. Jaramillo, M. *Perspectivas del Medio Ambiente Urbano; Technical Report: GEO Loja*; Programa de las Naciones Unidas para el Medio Ambiente (Oficina Regional para América Latina y el Caribe), la Municipalidad de Loja y Naturaleza y Cultura Internacional: Loja, Ecuador, 2007.
71. Emck, P. *A Climatology of South Ecuador—With Special Focus on the Major Andean Ridge as Atlantic-Pacific Climate Divide*; Friedrich-Alexander-University of Erlangen-Nürnberg: Erlangen, Bavaria, Germany, 2007.
72. Hungerbühler, D.; Steinmann, M.; Winkler, W.; Seward, D.; Egüez, A.; Peterson, D.E.; Helg, U.; Hammer, C. Neogene stratigraphy and Andean geodynamics of southern Ecuador. *Earth-Sci. Rev.* **2002**, *57*, 75–124. [CrossRef]
73. Soto, J.; Galve, J.P.; Palenzuela, J.A.; Azañón, J.M.; Tamay, J.; Irigaray, C. A multi-method approach for the characterization of landslides in an intramontane basin in the Andes (Loja, Ecuador). *Landslides* **2017**, *14*, 1929–1947. [CrossRef]
74. Schutt, B. Late Quaternary Environmental Change on the Iberian Peninsula. *Erde* **2005**, *136*, 3.
75. Trujillo, F., III. Clima e información meteorológica. In *PLAN INFOCA. Un Plan de Acción al Servicio del Monte Mediterráneo Andaluz*; Consejería de Medio Ambiente; Junta de Andalucía: Seville, Spain, 1995.
76. Gómez-Pugnaire, M.T.; Galindo-Zaldívar, J.; Rubatto, D.; González-Lodeiro, F.; López Sánchez-Vizcaíno, V.; Jabaloy, A. A reinterpretation of the Nevado-Filábride and Alpujárride Complexes (Betic Cordillera): Field,

petrography and U-Pb ages from orthogneisses (western Sierra Nevada, S Spain). *Schweiz. Mineral. Und Petrogr. Mitt.* **2004**, *84*, 303–322. [CrossRef]

77. Chacón, J.; Irigaray Fernández, C.; Fernández, T.; El Hamdouni, R. Landslides in the main urban areas of the Granada province, Andalucia, Spain. In *Engineering Geology for Tomorrow's Cities*; Engineering Geology Special Publication 22, IAEG2006 ed.; Culshaw, M., Ed.; The Geological Society of London: Nottingham, UK, 2006; p. 11.
78. Chacón, J.; Irigaray, C.; El Hamdouni, R.; Jiménez-Perálvarez, J. Diachroneity of Landslides. In *Geologically Active*; Williams, A.L., Pinches, G.M., Chin, C.Y., McMorran, T.J., Massey, C.I., Eds.; CRC Press Taylor & Francis Group: London, UK, 2010; Volume 1, pp. 999–1006.
79. Irigaray, C.; Lamas, F.; El Hamdouni, R.; Fernández, T.; Chacón, J. The Importance of the Precipitation and the Susceptibility of the Slopes for the Triggering of Landslides along the Roads. *Nat. Hazards* **2000**, *21*, 65–81. [CrossRef]
80. Palenzuela, J.A.; Marsella, M.; Nardinocchi, C.; Pérez, J.L.; Fernández, T.; Chacón, J.; Irigaray, C. Landslide detection and inventory by integrating LiDAR data in a GIS environment. *Landslides* **2014**, 1–16. [CrossRef]
81. Ferro Veiga, J.M. *Paisajismo, Iluminación y Decoración de Exteriores e Interior*; Createspace Independent Pub: Charleston, SC, USA, 2020; p. 371.
82. IDEAL. Historia. Available online: http://canales.ideal.es/acercaIdeal/historia.html (accessed on 15 September 2014).
83. NOAA.ESRL. Monthly Timeseries of El Niño 1+2 SST Index. Available online: https://www.esrl.noaa.gov/psd/gcos_wgsp/Timeseries/Data/nino12.long.anom.data (accessed on 15 December 2019).
84. NOAA.ESRL. Monthly NAO Normalized. Available online: https://www.cpc.ncep.noaa.gov/products/precip/CWlink/pna/norm.nao.monthly.b5001.current.ascii.table (accessed on 15 December 2019).
85. WeMO. Database and Resources. *WeMO*. Available online: http://www.ub.edu/gc/en/wemo/ (accessed on 15 December 2019).
86. NOAA. What are El Niño and La Niña? Available online: https://oceanservice.noaa.gov/facts/ninonina.html (accessed on 16 July 2020).
87. Ideo. North Atlantic Oscillation. Available online: https://www.ldeo.columbia.edu/res/pi/NAO/ (accessed on 16 July 2020).
88. Trauth, M.H.; Alonso, R.A.; Haselton, K.R.; Hermanns, R.L.; Strecker, M.R. Climate change and mass movements in the NW Argentine Andes. *Earth Planet. Sci. Lett.* **2000**, *179*, 243–256. [CrossRef]
89. Hoerling, M.P.; Kumar, A. Understanding and Predicting Extratropical Teleconnections Related to ENSO. In *El Niño and the Southern Oscillation: Multiscale Variability and Global and Regional Impacts*; Diaz, H.F., Markgraf, V., Eds.; Cambridge University Press: Cambridge, UK, 2000; pp. 57–88. [CrossRef]
90. Diaz, H.F.; Hoerling, M.P.; Eischeid, J.K. ENSO variability, teleconnections and climate change. *Int. J. Climatol. A J. R. Meteorol. Soc.* **2001**, *21*, 1845–1862. [CrossRef]
91. Can, R.; Kocaman, S.; Gokceoglu, C.A. Convolutional Neural Network Architecture for Auto-Detection of Landslide. Photographs to Assess Citizen Science and Volunteered Geographic Information Data Quality. *ISPRS Int. J. Geo-Inf.* **2019**, *8*, 300. [CrossRef]
92. Li, Y.; Sun, R.; Yin, K.; Xu, Y.; Chai, B.; Xiao, L. Forecasting of landslide displacements using a chaos theory based wavelet analysis-Volterra filter model. *Sci. Rep.* **2019**, *9*, 19853. [CrossRef]
93. Wu, C.L.; Chau, K.W. Prediction of rainfall time series using modular soft computingmethods. *Eng. Appl. Artif. Intell.* **2013**, *26*, 997–1007. [CrossRef]
94. Homsi, R.; Shiru, M.S.; Shahid, S.; Ismail, T.; Harun, S.B.; Al-Ansari, N.; Chau, K.-W.; Yaseen, Z.M. Precipitation projection using a CMIP5 GCM ensemble model: A regional investigation of Syria. *Eng. Appl. Comput. Fluid Mech.* **2020**, *14*, 90–106. [CrossRef]

Article

Comparing Forward Conditional Analysis and Forward Logistic Regression Methods in a Landslide Susceptibility Assessment: A Case Study in Sicily

Dario Costanzo [1,*] and Clemente Irigaray [2]

1 Department of Earth and Sea Sciences, University of Palermo, 90123 Palermo, Italy
2 Department of Civil Engineering, University of Granada, 18010 Granada, Spain; clemente@ugr.es
* Correspondence: costanzodario@gmail.com

Received: 15 June 2020; Accepted: 9 July 2020; Published: 10 July 2020

Abstract: Forward logistic regression and conditional analysis have been compared to assess landslide susceptibility across the whole territory of the Sicilian region (about 25,000 km^2) using previously existing data and a nested tiered approach. These approaches were aimed at singling out a statistical correlation between the spatial distribution of landslides that have affected the Sicilian region in the past, and a set of controlling factors: outcropping lithology, rainfall, landform classification, soil use, and steepness. The landslide inventory used the proposal of building the models like the official one obtained in the PAI (hydro geologic asset plan) project, amounting to more than 33,000 events. The 11 types featured in PAI were grouped into 4 macro-typologies, depending on the inherent conditions believed to generate various kinds of failures and their kinematic evolution. The study has confirmed that it is possible to carry out a regional landslide susceptibility assessment based solely on existing data (i.e., factor maps and the landslide archive), saving a considerable amount of time and money. For scarp landslides, where the selected factors (steepness, landform classification, and lithology) are more discriminate, models show excellent performance: areas under receiver operating characteristic (ROC) (AUCs) average > 0.9, while hillslope landslide results are highly satisfactory (average AUCs of about 0.8). The stochastic approach makes it possible to classify the Sicilian territory depending on its propensity to landslides in order to identify those municipalities which are most susceptible at this level of study, and are potentially worthy of more specific studies, as required by European-level protocols.

Keywords: landslide susceptibility assessment; forward logistic regression; forward conditional analysis; GIS; Sicily

---

1. Introduction

One of the most obvious effects of rapid territorial expansion in recent decades is the growing impact that natural disasters have on man and his activities. Therefore, institutions are committed to investing their resources in both the implementation of structural interventions to mitigate risk as well as early warning systems, and in defining guidelines for land management and civil protection issues [1,2]. Landslides are among the major contributors to the dynamics of the morphological evolution of slopes and occur when slope stability conditions change due to increased stress or decreased resistance along a failure surface. A deformation occurring on a pre-existing failure surface is called re-activation, whilst one along a new fracture plane is referred to as a neo-activation. International literature refers to landslide susceptibility as the spatial probability for gravitational instability conditions within an area, based on its physical-environmental conditions [3,4]. Therefore, depending on the spatial variability of the physical-environmental features of the study area (typically, a river-basin or an administrative territorial unit), a landslide susceptibility map allows the units into which it is subdivided to be differentiated, according to the higher or lower probability of a landslide occurring. Most approaches and methods were designed to evaluate landslide susceptibility based on the identification and spatial characterization of a set of control factors, and on quantifying the relationships between these and an archive of past landslides. This concept is a reinterpretation of the actualism principle [5] from a geomorphological point of view: the past and the present are the keys to the future [1,6–8]. This principle suggests that the areas affected by landslides in the future will be those that share similar characteristics to those already recorded in past landslides [7,8]. One of the first aspects to be faced in planning research stages aiming to define a set of susceptibility conditions of an area is to establish the scale and approach at which the analysis should be performed. Time- and economic resources-allowing geostatistical approaches are often the most suitable for areas of hundreds or thousands of square kilometers [4,9]. Landslide susceptibility assessment poses specific methodological issues when performed for regional mapping purposes [2]. Indeed, within regional applications, overall forecasting performance is seriously hindered by the lack of data required or the inaccuracy of these data: landslide inventories and thematic maps of the controlling factors. Likewise, no matter what the resolution of the processed data is, some basic issues in model building procedures, such as the modelling approach, mapping units, landslide classification, and representation and validation strategies [1], need to be optimized when applied to regional multi-scale assessment procedures. Thus, an expert European landslide group has recently proposed some criteria for a multi-level method (TIER [10]) in order to define shared approaches to landslide susceptibility mapping. Three susceptibility TIER levels have been proposed and reference data and model building procedures have been recommended for each one. The TIER approach is strictly dependent on the quality of the available data in landslide inventories and on thematic maps, required for the whole European territory. This protocol consists of three nested levels (TIER1, TIER2, and TIER3) with a gradually increasing degree of resolution for the predictive models [10–13]. In a nested tiered approach, when the scale of work changes, the resolution of the factors, the mapping units, as well as the complexity and the type of the techniques used, may change too. The TIERS protocol features a high degree of objectivity and spatial and/or time repeatability, which is why it represents an important safety benchmark for national and regional administrations. For small-scale studies (smaller than 1:100,000), the methods recommended by the Joint Research Centre (JRC research group) are those based on expert-driven approaches such as conditional analysis, heuristics, and/or weighted factors. In this piece of research, we have had the opportunity to verify and compare the use of the binary logistic regression (BLR) statistical technique on extensive areas, in the range of tens of thousands of square kilometers, contrary to suggestions by European Commission experts (Tier-based approach). We have then compared results with those obtained through the conditional analysis (CA) approach and with those obtained by other authors researching the same area [10,13,14]. Susceptibility scenarios described by the maps are also compared and homogenized with those hazards arising from the PAI program [15], working towards a punctual and detailed analysis of all possible discrepancies that may result. Further to validation, and once the robustness of the scientific guidelines testing has been tested, the skills and the experience acquired may be used as a basis for the drafting of a Sicilian municipality susceptibility map, a useful tool for the policy-makers dealing with land management.

2. Study Area

The geological setting of Sicily (Figure 1) consists of three main structural elements [16,17]: a range sector, running along the northern strip of the island, from the Peloritan Peaks to the Aegadian Islands, where Triassic to Mesozoic structural–stratigraphic units, mainly characterized by carbonate (at the base) to clayey (at the top) formations, tectonically overlap; a northwest-dipping foredeep area lying in the mid-southern area of the Sicilian territory, consisting of Plio-Pleistocene pelagic marly limestones, silty mudstones, and sandy clays overlying Messinian evaporites separated by the Sikan Peaks; the Hyblean Plateau located in southeastern Sicily, made up of a Triassic-Liassic platform and scarp-basin carbonates overlaid by Jurassic-Eocene pelagic carbonates and Tertiary open-shelf clastic deposits;

southern and central Sicily feature Cretaceous-Lower Pleistocene clastic-terrigenous deposits and Messinian evaporites [18].

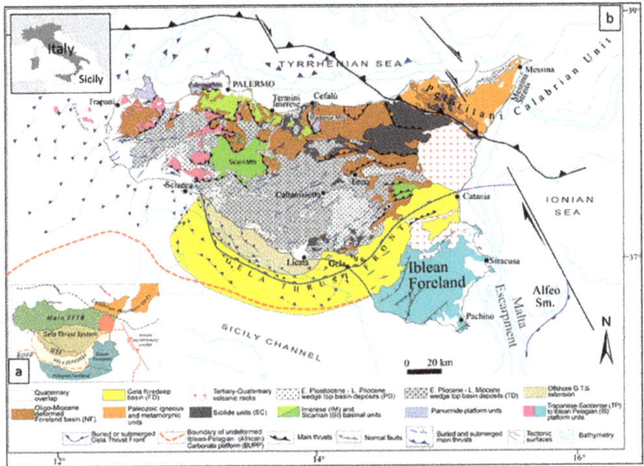

Figure 1. Schematic structural map of Sicily [17]. (**a**) The three main geological structural elements of Sicily; (**b**) distribution of the main stratigraphic–structural units.

The mechanical characteristics of outcropping lithotypes are among the main geo-environmental factors directly influencing the geomorphological stability of Sicilian slopes. Slides and flows are mainly located where the clay continental sequences outcrop. Hard block outcrops (metamorphic and carbonate) are affected by falls, topples, and lateral spreading. The rainfall may cause denudation slope actions able to generate, in the presence of debris on the metamorphic units, debris flows or debris avalanches. The triggering factors which are most likely to influence the activation of landslides in Sicily, are, in order: rainfall, human activity, volcanic eruptions, and earthquakes. Carbonatic rocks outcropping in the foreland sector are almost exclusively affected by falls. Both the ductile clayey formations and the weathered top coverage of the metamorphic units may experience rapid debris avalanche/debris flow phenomena.

The geomorphological features of the island of Sicily have been crafted by the collision of the Eurasian and African plates acting in synergy to shape the current landscape. The topography is directly influenced by the stratigraphic structure of the area and by the surface uplifting and subduction that occurred during the quaternary period, influenced heavily by significant eustatic sea level variations. The changes to Sicily's geology through the epochs have led to a mountainous and hilly terrain.

Furthermore, the influence of medieval human civilization is still apparent in the small Sicilian urban centers. These historic centers are often surrounded by harsh terrain, where slopes with a gradient of 50% or more can often be seen encompassing the centers, features which would have been favorable to the population who first settled there when defending their land. This, however, leads to limited space and opportunity for further urban development.

The flat areas of the island, a total of just 7% of the entire territory, are represented by the alluvial plain of Catania, the coastal plain of Licata and Gela, the coastal area of Trapanese, and that between Syracuse and Scicli, at the foot of the Monti Iblei.

From analysis of the rainfall data of the Sicilian stations, it is possible to highlight how the rainfall is concentrated, especially in the October–March period, though it is somewhat appreciable in the spring (April–May) and of little importance in the summer months. The maximum and minimum values of average annual precipitation are respectively 919 mm in the Monreale station, in the province of Palermo, and 510.6 mm in the Agrigento station. Frequently, at the highest peaks, there is snowfall, of which there is no significant long-term data, due largely to the lack of a sufficient number of snow stations. A meteoric event of considerable importance in the northern mountain ranges, in particular for the northern slopes of the Madonie due to the presence of fog, which, in addition to integrating normal water supplies through condensation, performs a mitigating and compensating action for extreme climatic events, limiting precipitation and keeping temperatures lower during periods of summer water deficit, as well as decreasing the intensity of weather events harmful to plants, such as late frosts. As regards the thermometric data, there is an inverse trend compared with that of rainfall, as occurs throughout the Mediterranean region. There is, in fact, a gradual increase between March and April, and a more marked increase from May to July–August—a period in which the absolute maximum values are reached—beyond which the temperatures gradually decrease until October, and then drop sharply until December and touch the minimum values in January–February which is the coldest time of the year. The highest average annual temperature is 18.8 °C for the Cefalù (Palermo) station, while the lowest, 13.3 °C, is recorded in Petralia Sottana (Palermo). The lowest average annual minimum temperature value (9.3 °C) is recorded in the Petralia Sottana station, while the highest annual average maximum temperature value (24.1 °C) in the Lentini (Siracusa) and Palermo stations (Castelnuovo Institute).

3. Data Collection and Processing

3.1. Landslide Inventory

Landslides are natural events in the evolution of a slope. They are a problem and become a hazard and/or risk when they interact with man and the man-made environment. Landslides can be classified according to their movement types and the nature of the displaced material type, as well as the state, distribution, and style of their activity [8,19–21]. As raised by [22], there is a conceptual ambiguity concerning landslides stemming from the use of the very same term (i.e., landslide) for both the landslide deposit (displacement volume) and the movement of material on a slope or a pre-existing landslide body [20,23]. This is in addition to general confusion originating from the variable and complex nature of the phenomenon itself [24], due to profoundly different morphological characteristics, behavior, activity states, and their evolution. The construction of the landslide inventory is a fundamental and critical step towards the application of statistical models designed to estimate the probability of new activations affecting previously uninvestigated areas. A landslide inventory commonly represents the sum of all the events that have occurred in an area. Alterations to a slope profile, pointing to ongoing landslides, tend, over time, to become less evident because of erosion, new landslides, human activity, and vegetation, making the "in landslide/not in landslide" border hard to detect with the passage of time. Generally, "newer" phenomena generated by recent heavy rainfall or earthquakes are more easily identifiable and interpretable than more remote ones, where diagnostic elements begin to dissolve. Certainly, there is a critical issue concerning the updating of the landslide inventory, as the public administration lacks economic and human resources. This shows, even more clearly, the limitations of PAI methodology in mapping areas at risk, as it leads to a zoning system, which is not closely related to all new activations. No systematic archive of slope instability exists for the research area of this paper. The latest archive of slope failures is the one belonging to the PAI project, which now holds 33,094 landslides (latest update May 2016) classified into 11 different types. The landslides that were mapped within the PAI project affect an area of approximately 1300 square kilometers, approximately 5.1% of the size of the region (Table 1; Figure 2). The landslide archive derives from a historical inventory and territorial analysis, completed with the execution of inspections, which began in 2003. The archive is continuously updated by reports from the following: literature and scientific publications; studies to support urban development projects in municipalities; regional civil protection archives; and reports from local authorities to regional and national bodies (Civil Protection, Territory and Environment Departments) of geomorphological phenomena that have occurred (from 1998 to today).

Table 1. Landslide inventory, statistics, and their reclassification.

PAI		Area [m²] for a Single Landslide					Percentages		SUFRA		
TYPOLOGY	Number of Cases	max	min	mean	Std. Dev.	Total (km²)	Landslide Area	Total Area	Number of Cases	Area (km²)	TYPOLOGY
1. Falls/Topples	5460	1,630,259	25	15,310	1,138,586	84	6.4	0.3	5460	84	1. Scarp Landslide
2. Rapid flows	853	1,152,779	79	15,403	48,938	13	1.0	0.1	10,202	435	2. Hillslope landslide
4. Slides	2835	45,106	530	11,350	11,987	81	6.2	0.3			
5. Complex	3076	735,524	766	50,290	114,782	215	16.5	0.8			
7. Slow flow	3438	194,381	869	34,263	41,727	125	9.6	0.5			
6. DPGV/Spreads	28	7,891,513	212	509,802	509,802	14	1.1	0.1	28	14	3. DPGV and Spreads
3. Sinkhole	43	1,380,425	25	51,918	219,976	2	0.2	0.0	17,404	772	4. Others
8. Areas with diffused landslides	2877	225,203	1482	53,560	225,203	288	22.1	1.1			
9. Slowly surface deformation	3512	124,610	605	23,534	29,059	155	11.9	0.6			
10. Badlands	1266	1,625,318	3471	42,755	302,626	54	4.1	0.2			
11. Caused by accel. erosion	9706	254,775	9358	30,171	57,136	273	20.9	1.1			
Total	33,094				2,699,822	1305	100.0	5.1	33,094	1305	Total
Area of Sicily 25.832 Km²											

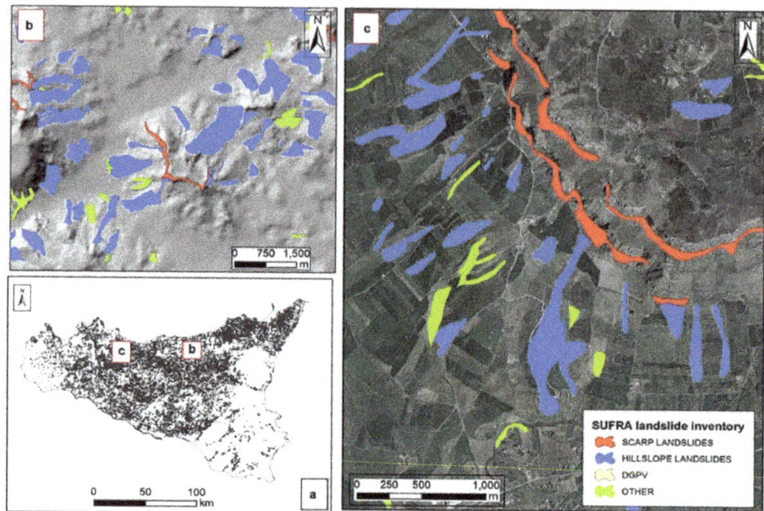

Figure 2. PAI landslide inventory. (**a**) Regional distribution of landslides; (**b**) detail of a portion of the territory of landslides on the hillshade grid; (**c**) PAI landslides on orthophotos.

The PAI archive focuses more on urban and populated areas. In this piece of research, various models were created to record landslides affecting scarps (scarp landslides; SCR_LSN) and landslides lying over slopes (hillslope landslides; HILL_LSN).

3.2. Modelling Approach

A predictive model must represent the response of a natural system to the trigger conditions described by its environmental features; the response may lie in the spatial distribution of new landslides or in the so-called prediction image. The effectiveness of the model can be measured by comparing the final expected results (the susceptibility map) with the actual results that are directly observed (the new landslide map). When studying the natural environment, the use of models is essential, as they allow for a simplification of the infinite natural variables, as well as operating within an acceptable processing time with conventional computers. Thanks to a greater exchange of information, the development of the hardware and software component for the acquisition and processing of data, and the interaction between different research groups, the methods for assessing susceptibility from landslides have evolved rapidly [3,4,24–27]. In the last decade, several applications have been carried out with the aim of comparing results from different statistical approaches, as applied to the implementation of models of landslide susceptibility on the same area, using the same landslide inventory and the same control factors [28–31]. Although full agreement does not exist within the scientific community as to what the best approach to follow is, the experience gained through our previous studies tells us which statistical techniques may be adopted from among the many available (discriminant analysis, conditional analysis, binary logistic regression, classification, and regression trees) and which one leads to greater performance in terms of predictive fitting, and robustness [27,32–39].

3.3. Conditional Analysis (CA)

The CA statistical approach has been widely adopted in landslide susceptibility assessment [24,26,33,40–44]. The CA concept is simple and easily manageable in a GIS environment. That is why it is the single most recommended method when investigating the landslide susceptibility conditions of larger areas and at scales greater than 1:100,000.

The CA method is based on the concept that landslide density, computed on homogenous domains, represents the relational function between the environmental variables and the framework of a landslide area [2,26,45].

According to the CA concept, we must compute the density of cells in an unstable area for unique condition units (UCUs), defined by overlapping and combining a set of selected control factors and intersecting the UCU and landslide layers. From a statistical point of view, the landslide density corresponds to the susceptibility value of an area in a landslide, linked to a particular combination of control factors. Mathematically, this concept can be expressed as

$$P = \delta UCU(i) = \frac{UCUunst(i)}{UCU(i)} = \frac{\left(\frac{UCUunst(i)}{UCUunst}\right) * \left(\frac{UCUunst}{UCUall}\right)}{\left(\frac{UCU(i)}{UCUall}\right)} \quad (1)$$

where probability (P) can be computed as the ratio between unstable (UCUunst) and total counts (UCUall) of cells for the cells which have a value.

Landslide density is the relational function between the environmental conditions and the landslides in a given area, meaning that a density ranking order corresponds to a scale of landslide susceptibility [32,33,46].

3.4. Binary Logistic Regression (BLR)

One of the key points in determining the susceptibility conditions of an area with multivariate statistical techniques is the selection of an appropriate number of factors that can justify the spatial distribution of past and future forms of instability. In fact, many of these techniques provide an estimate of the importance of each factor in relation to the others, or its specific contribution in generating a particular type of landslide in the area under investigation. Many of these techniques rank, through hierarchization, the contribution of each factor in determining the landslide-specificity of an area by identifying the minimum and maximum number of factors needed, beyond which the performance variation of the model may be defined as insignificant or even negative [47,48].

According to Hosmer and Lemeshow [49], the aim of an approach based on binary logistic regression (BLR) is to enable the singling out of the best linear relationship between a dichotomous dependent variable (such as 1 or 0 representing the "presence"/"absence" of landslides, respectively) and a set of independent variables representing control geo-environmental factors. In the logistic regression equation, the expected dependent variable f(x) may be expressed as

$$\text{logit}(x) = \ln(\text{odds}) = \ln\left[\frac{\pi}{1-\pi}\right] = \alpha + \beta_1 X_1 + \beta_2 X_2 + \cdots + \beta_p X_p \quad (2)$$

where logit(x) corresponds to a natural logarithm of odds $[\pi/(1-\pi)]$, expressed as a ratio between the likelihood of the presence of landslides (π) over the likelihood of their absence $(1-\pi)$; α is the intercept of the model; and β_1, β_2 up to β_n are the coefficients, which measure the contribution of each independent input variable [50–52].

In other words, the BLR allows us to estimate the contribution of each input variable using its coefficient in the probability equation by adopting the maximum likelihood classifier concept. Comparing the maximum likelihood computed for every β-value with each estimated error, the significance of the coefficients is tested using the Wald test [31,49,53]. The probability of occurrence can be estimated by multiplying the −2 log-likelihood ratio; thus, the negative log-likelihood (−2LL) statistic is obtained, which has a chi-square ($\chi 2$) distribution. According to this approach, BLR is executed through a stepwise procedure that allows the differentiation of only those predictor variables with a significant impact on the performance of the multivariate model [36,54,55].

A problem arising when using BLR is that which is related to choosing the appropriate sample size for an unstable area (positive cases) as opposed to stable ones (negative cases). Indeed, the number of positive cases is significantly lower than negative ones, thus generating an imbalance. Therefore, it is one of the most important choices concerning the random selection technique likely to lead to a balanced dataset of landslide and non-landslide pixels [56]. In this piece of research, the TANAGRA open-source software was used to apply forward stepwise logistic regression [57].

3.5. Variables Selection and Factor Class Definition

Landslides are directly connected to the many environmental condition types, mainly depending on slope morphology (slope angle, orientation slope, curvature, elevation, roughness, etc.) and other intrinsic characteristics (lithology, soil use, tectonic condition, landform classification, etc.), while the activation of new landslides depends on trigger factors, such as intensive rainfall or earthquakes [58–60].

All kinds of geo-environmental factors may be considered as potential predictive factors and can be introduced into the analysis. All the factors should be drawn at a specific measurement for classification (categorical, continuous). Acquiring each factor requires time and money. In addition, we must also consider the computing time needed to process huge volumes of data. This is why only those variables thought to be directly and/or indirectly capable of conditioning the established of a slope, thanks to knowledge acquired from previous studies, may enter the analysis. The use and reclassification of a limited number of factors also prevents the generation of a large number of not very widespread mapping units and thus the overestimation of the density for untrained mapping units.

The CA method, requires preselecting the factors to be entered into the analysis, and then defining the UCUs representing the mapping unit in this type of analysis. The univariate approach was followed for each acquired variable and their density calculated for the different types of landslide analyzed (scarp and hillslope landslide). The univariate approach was followed for each acquired variable and their density calculated for the different types of landslide analyzed (scarp and hillslope landslide). Univariate analysis allows to distinguish the significant from insignificant variables, and then combine only those variables among them which are likely to be the most influential in the production of the mapping units GRID layer. The following continuous and categorical variables have been acquired and analyzed (Table 2).

Table 2. Geo-environmental factors and their spatial distribution.

Categorical Variables				
Variable	**References**	**Description**	**Code**	**Percentage Distribution (%)**
Soil Use	(Corine Land Cover project, 2006)	Continuous urban fabric	USE_111	1.94
		Discontinous urban fabric	USE_112	2.42
		Transitional areas	USE_13	0.27
		Green urban areas	USE_14	0.06
		Arable land	USE_21	32.54
		Permanent crops	USE_22	21.61
		Heterogeneous agricultural areas	USE_23	14.97
		Forest	USE_31	7.79
		Shrub and/or herbaceous associations	USE_32	17.31
		Open spaces with little or no vegetation	USE_33	0.65
		Water bodies	USE_51	0.44

Table 2. *Cont.*

Categorical Variables				
Variable	**References**	**Description**	**Code**	**Percentage Distribution (%)**
Outcropping lithology	Lithological Complex	Continental clastic deposition complex	LITH_CDC	12.94
		Phyllitic and metamorphic complex	LITH_PhMe	3.51
		Sandy-calcarenitic complex	LITH_SaCa	13.22
		Evapotitic complex	LITH_Ev	4.86
		Conglomerate-sandstone	LITH_CoSa	2.74
		Clay complex	LITH_Cl	34.13
		Sandstone and clay	LITH_SaCl	8.66
		Carbonatic complex	LITH_Ca	13.41
		Volcanic complex	LITH_Vo	6.53
Landform classification	Landform Classification (Weiss, 2001)	Canyons	LCL_CANY	7.51
		Midslope drainage	LCL_MDRG	2.27
		Upland drainage	LCL_UPDRN	2.32
		U-shaped valleys	LCL_USHP	2.92
		Plains	LCL_PLAINS	32.58
		Open slopes	LCL_OPEN	39.51
		Upper slope	LCL_UPPSL	2.80
		Local ridge	LCL_LOCRDG	0.00
		Midslope ridge	LCL_MRDG	2.00
		Mountain tops	LCL_MNTPS	8.07
Rainfall (mm)	SIAS, 2015	0–450	RAIN_L	1.58
		450–600	RAIN_M	63.32
		600–800	RAIN_H	20.75
		>800	RAIN_VH	14.34
Slope Angle (Scarp landslide)	Θ = TAN Δy/Δx	Canyons	SLO_L	87.81
		Midslope drainage	SLO_M	10.72
		Upland drainage	SLO_H	1.47
Slope Angle (Hillslope landslide)		U-shaped valleys	SLO_L	78.09
		Plains	SLO_M	15.28
		Open slopes	SLO_H	6.43
		Upper slope	SLO_VH	0.20

3.5.1. Continuous Variables

The 2-m ARTA-DEM was used and resampled to generate the digital elevation model (DEM). Slope angle, slope aspect, and landform classification were calculated using the DEM.

- Slope angle (SLO) is usually considered as one of the main controlling factors in landslide modelling. At first, SLO was classified into 5 natural break intervals [14], expressed in sexagesimal degrees (0°–5°; 5°–12°; 12°–18°–18°–32°: > 32°). The raster-file of the slope angle was obtained by resampling the 2-meter resolution ARTA-DTM flight ATA (2007/2009) to 100 m per side. As shown in Figure 3, the proposed reclassification for the slope angle for the hillslope landslide does not reveal the theoretical concept for the slope increase, which corresponds to an increase in the likelihood of landslides occurring. This does not happen with the scarp landslide, where increasing the slope angle leads to an increase in the percentage of landslides.

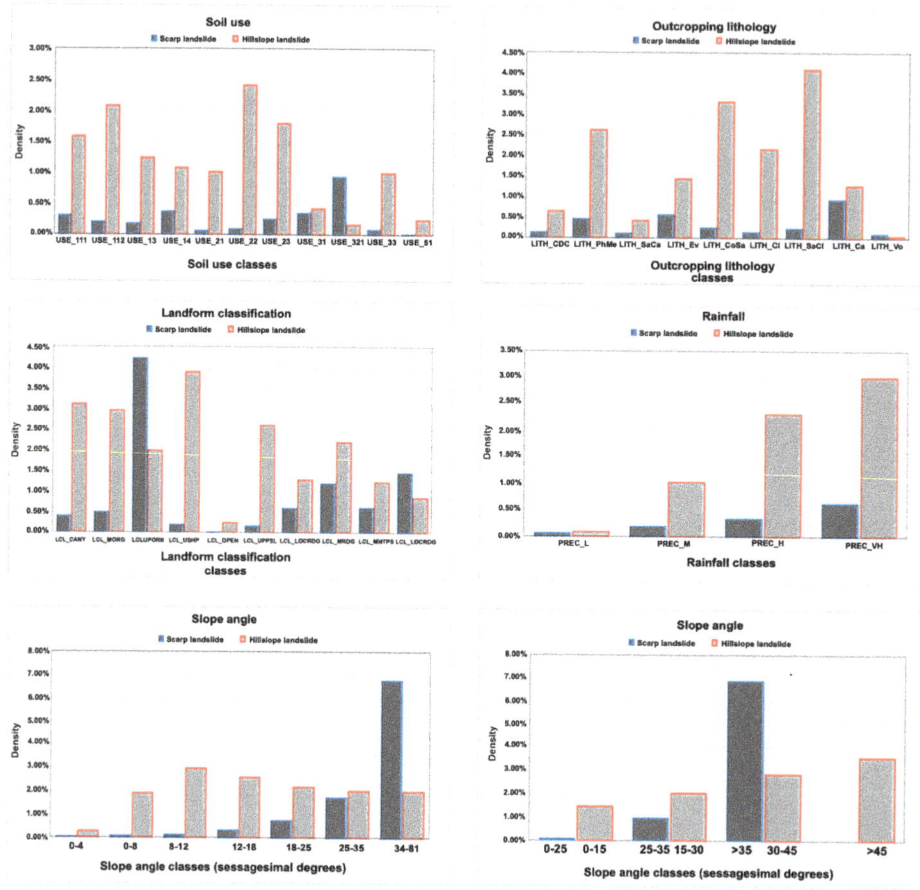

Figure 3. Univariate analysis for the continuous variables. **Soil use**. USE_111: continuous urban fabric; USE_112: discontinuous urban fabric; USE_13: transitional areas; USE_14: green urban areas; USE_21: arable land; USE_22: permanent crops; USE_23: heterogeneous agricultural areas; USE_31: USE_321: shrub and/or herbaceous associations; USE_33: open spaces with little or no vegetation; USE_51: water bodies. **Landform classification.** LCL_CANY: canyons; LCL_MDRG: midslope drainage; LCL_UPDRN: upland drainage; LCL_USHP: U-shaped valleys; LCL_PLAINS: plains; LCL_OPEN: open slopes; LCL_UPPSL: upper slope; LCL_LOCRDG: local ridge; LCL_MRDG: midslope ridge; LCL_MNTPS: mountain tops. **Rainfall (mm).** PREC_L: 0–450; PREC_M: 450–600; PREC_H: 600–800; PREC_VH: >800. **Slope Angle** (scarp landslide). SLO_L: 0°–25°; SLO_M: 25°–35°; SLO_H: >35°; **Slope Angle** (hillslope landslide). SLO_L: 0°–15°; SLO_M: 15°–30°; SLO_H: 30°–45°; SLO_VH: >45°.

3.5.2. Category Variables

- Landform classification (LCL). Using an ArcMap open source tool, the LCL variable was derived directly from the DEM. LCL provides a simple and repeatable method to classify the landscape into slope position and landform category comparison. The different landform category classes can be determined by classifying the combination of a small and large neighborhood topographic position index (TPI) computed for each cell from different scales. The TPI is simply the difference between a cell elevation value and the average elevation of the neighborhood around that cell. Positive values mean the cell is higher than its surroundings, while negative values mean it is lower [61]. To compute the LCL, the small and the large neighborhood areas were set to 500 and 100 m, respectively. Ten landform classes were thus obtained (Table 2);

- Outcropping lithology (LITH). Together with the slope itself, the lithological conditions of an area are the most important factors influencing the geomorphological processes on the slope. The lithology controls the response of the slope in terms of the trigger-time of the collapse because of rainfall or seismic forces and evolution of the process. The lithotypes cropping out in the map of the Sicilian region were used in this research and grouped into 9 different "lithological complexes", according to their geotechnical characteristics. The output lithological complexes were named as shown in Table 2. The clay complex is the most widespread one in the Sicilian territory, as it crops out in almost 35% of the area (more than 50,000 hectares);
- Soil use (USE). In this test, we used a soil use map derived from the 1:100,000 Corine Land Cover project (2006) based on a revised version of the Corine Land Cover 2000 dataset with the results of Landsat 1988 and photointerpretation of aerial photos. Table 2 shows land cover characteristics in 11 different classes, for terrain units larger than 0.25 km^2. The Corine 2006 map was converted into a soil cover digital map provided by the Sicilian region, using the second level of the Corine legend, except for the "urban areas" class, which has been divided into continuous (USE_111) and discontinuous (USE_112) urban fabric, corresponding to level III of the Corine Land Cover classification. Arable land (USE_21) covers more than 30% of the research area. Forest crops cover about 7% of the area and mainly appear in the northeastern sectors. Areas covered by shrubby and herbaceous vegetation associations are dispersed around the study area: they cover 17%. Urban area cover is only 4.5%;
- Rainfall (RAIN). For rainfall, 280 rainfall stations were used to create the GRID rainfall map using the inverse distance weighted method. The database from the Sicilian regional administration office (http://www.osservatoriodelleacque.it) was used to extract the mean annual precipitation (for the period 1921–2009). Regarding precipitation, Sicily can be divided into three main sectors with three different pluviometric regimes: the northern sector: includes all the Tyrrhenian coast of the island. Rainfall here is characterized by a rainy season (autumn–winter) and a dry spring and summer. Eastern Sicily: in this area, rainfall is also greater in winter. Precipitation is often concentrated into short spells and is sometimes very violent. This is because the precipitation depression bearers come from Africa and are very hot and humid, favoring strong thermal contrasts. Southern Sicily: includes all the area bordered by the Mediterranean Sea. As in the rest of the island, winter is the rainy season. The number of rainy days is less than in the northern area (<60 days per year). In some areas, rainfall is sparse, especially in the coastal zone. The areas with the highest rainfall are the Madonie, Nebrodi, and Peloritani peaks, Etna, and the area south of Palermo. The driest areas are the Plain of Catania and the southern coast, in particular, Gela city.

Generally, there are two reasons that lead us to choose the smallest possible number of geo-environmental variables for the construction of the forecasting model, which allow the realization of what is called the "best model" for a specific area, capable of providing an acceptable performance forecast. On the one hand, achieving or obtaining each parameter requires spending a considerable amount of time and money, on the other, a large number of environmental variables results in a large number of possible combinations characterizing each of the territorial units chosen in an excessively specific manner, as the basis for statistical analysis. A high number of combinations bring a progressive decrease in the distribution of each combination class. The consequence is an unexpected decrease in the performance of the susceptibility model caused by the inclusion of variables which are closely related to a small number of cells, but poorly correlated to the global distribution of the remaining part, thus affecting the choice of the most predictive variables. Even the selection of factors is an essential step in landslide susceptibility assessment procedures, in which the nature of geomorphological criteria takes priority.

An expert-driven univariate analysis procedure for the following 5 processing control variables was carried out (Figure 3). Depending on the type of landslides, a maximum moderation criterion in the number of factors used is necessary to identify a first set of control factors that can be justified based on morphodynamic models, defined as heuristic at first approximation, for the distribution of the observed phenomena. Then, regression techniques may be applied to highlight the actual role played by each of the geo-environmental variables considered. However, it is also common practice, and recommended by applied statistics handbooks, to maintain certain diagnostic values of the variables (i.e., slope), even when stepwise regression procedures have greatly reduced their influence. A type of approach which, in a certain way, is the opposite of the analytical-geomorphological way, relying on deterministic or physically based methods.

That is why, for example, for "Slope angle", values were reclassified into 4 different classes (for hillslope landslide) and 3 classes (landslide scarp). At this point, a new univariate analysis was carried out and the results are shown in Figure 3.

4. Model-Building Strategy

The five reclassified geo-environmental variables were combined in unique conditions units (UCUs) aiming to get all the possible combinations of the classes of the different factors in order to assess the landslide susceptibility. The UCU layers were laid over with the landslide ones (SCR_LSN and HILL_LSN, respectively). The landslide densities were computed for each UCU class combination, as the ratio between landslide area counts and total pixel counts: according to Bayes' theorem [26,45], these values express the conditional probability of landslide occurrence, given a factor condition. The variables were combined and a UCU layer was derived from these. The combination produced a large number of combination classes (8355 for the HILL_LSN; 4173 for the SCR_LSN). The CA method requires that the combination of output factors has total areas which are large enough to guarantee the statistical significance of the "observed" sample. Table 3 shows the combination classes of the most susceptible UCUs. The UCUs found to have the highest susceptibility values are those characterized by predominantly consistent lithologies in SCR_LSN, (LITH_Ca), LCL_MNTPS (as landform classification), and high and very high slope angle values. Though uncommon, these UCU combinations have a susceptibility value of 100%. For hillslope landslides, clayey lithologies and open space LCLs are the most widespread combinations capable of generating slope failures (Table 3).

Table 3. Most susceptible unique conditions units (UCUs) in the target area.

Most diffused UCUs for SCR_LSN							
UCU Code	**Area (Ha)**	**LCL**	**LITH**	**USE**	**SLO**	**RAIN**	**δ**
1502	2	LCL_MNTPS	LITH_Ca	USE_112	SLO_VH	RAIN_VH	100.00%
1515	3	LCL_UPDRN	LITH_Ca	USE_23	SLO_VH	RAIN_VH	100.00%
1204	8	LCL_USHP	LITH_SaCa	USE_13	SLO_H	RAIN_H	100.00%
1517	3	LCL_MNTPS	LITH_Ca	USE_13	SLO_H	RAIN_L	100.00%
1217	3	LCL_MNTPS	LITH_Ca	USE_13	SLO_VH	RAIN_VH	100.00%
1001	2	LCL_UPDRN	LITH_SaCa	USE_13	SLO_H	RAIN_VH	100.00%
4095	1	LCL_USHP	LITH_CoSa	USE_13	SLO_M	RAIN_H	100.00%
1075	2	LCL_UPDRN	LITH_Ca	USE_13	SLO_VH	RAIN_M	100.00%
3403	1	LCL_USHP	LITH_Ca	USE_13	SLO_VH	RAIN_M	100.00%
3850	3	LCL_UPDRN	LITH_CoSa	USE_112	SLO_H	RAIN_VH	100.00%
Most diffused UCUs for HILL_LSN							
UCU Code	**Area (Ha)**	**LCL**	**LITH**	**USE**	**SLO**	**RAIN**	**δ**
2778	14	LCL_OPEN	LITH_CI	USE_13	SLO_VH	RAIN_VH	100.00%
3230	11	LCL_UPDRN	LITH_CI	USE_21	SLO_H	RAIN_H	89.00%
2759	34	LCL_OPEN	LITH_CI	USE_21	SLO_L	RAIN_M	62.00%
2777	17	LCL_OPEN	LITH_CI	USE_13	SLO_VH	RAIN_M	57.00%
3370	13	LCL_UPDRN	LITH_CI	USE_32	SLO_VH	RAIN_VH	36.00%
2711	45	LCL_OPEN	LITH_CI	USE_13	SLO_H	RAIN_VH	31.00%
4376	14	LCL_OPEN	LITH_SaCa	USE_21	SLO_H	RAIN_H	25.00%
2735	32	LCL_OPEN	LITH_CI	USE_13	SLO_M	RAIN_L	24.00%
1272	14	LCL_UPDRN	LITH_SaCa	USE_21	SLO_M	RAIN_VH	24.00%
3381	7	LCL_USHP	LITH_CI	USE_32	SLO_VH	RAIN_VH	23.00%

5. Results and Validation

Regardless of the statistical approach, models and susceptibility maps should always be validated to test the forecasting performance of the proposed protocol. Generally, a statistical validation procedure correlates the performance forecast (the expected landslides) that represents the space distribution of the model created by using a part of the observed phenomena (training dataset) with the distribution of a set of landslides which are not used in model-building (test dataset). This test dataset should therefore be temporally or spatially different from the training dataset used for model building, but the regional administration does not provide a systematic and continuous mapping of new landslides which occur reasonably often and in all Italian regions and worldwide. Many validation approaches have been presented in the last two decades in the relevant literature. In this paper, the random partition strategy was used, based on the random partition of the slope failures in two balanced populated training and test subsets to compare and estimate the robustness and reliability of the susceptibility models proposed.

The models presented in this research were derived using the CA and BLR statistical approaches, as the result of the average of 100 different replicas obtained by splitting the dependent variable, in 75% TRN_Subset and 25% TST_Subset randomly 100 times.

A reliable approach to test the performance of susceptibility models is the receiver operating characteristic (ROC) curve, allowing the comparison of the predictive performance models. The ROC curve is a plot of the probability expressing the sensitivity (TP rate) that represents the area classified correctly as susceptible (x-axis) versus the 1-specificity (FP rate), representing the probability of false prediction in response to an event for all the cutoff probability values (y-axis). ROC curve analysis allows the differentiation between two classes of events: unstable and stable cells [62]. The quantitative measure of model performance can be tested by computing the area under the curve (AUC) ranging from 0 to 1 [63]. The closer the AUC values are to 1, the higher the predictive performance of the model will be, while the closer the values are to 0.5 (random performance) the higher the inaccuracy of the model will be [50,64,65]. A value equal to 1 denotes a perfect discrimination between positive and negative cases. ROC curves were created for each of the two different statistical approaches (CA and BLR). In Figure 4, the ROC curves for both CA and BLR models were drawn for the training and test subsets with the aim of evaluating the model fitting for both approaches.

The BLR approach has been applied for forward stepwise selection independent predictive variables [28]. As shown in Table 4, average results for the 100 different splits of dependent variables in terms of SUCU. The model suite produced for SCR_LSN is characterized by a mean error rate of 0.17 (St.dev. 0.004) and AUC > 0.9 (outstanding). The graphs (Figure 4) show the average AUC values are close to excellent for HILL_LSN (>0.77) according to Hosmer and Lemeshow [49].

For HILL_LSN, the mean error rate is higher (about 0.29) but still very stable (St. dev. 0.001): a ranking of predictor variables derived from exploiting the forward stepwise statistical procedure.

Using BLR grid-cell-based models, the positive cases (unstable cells) are dramatically less than the negative cases (stable cells) and a suite of 10 different models (both for SCR_LSN and HILL_LSN) were prepared in order to estimate the model fitting, prediction skill, and robustness of the proposed approach [22,28,30,36,66]. Each suite model is made up of a subset of the unstable/positive cells and by an ever-changing subset which contains the stable/negative cells. The SCR_LSN grid-cell models were prepared by merging 6674 positive cells (5% of the total area) with 10 different subset, randomly selected negative 100 × 100 m cells. As for the HILL_LSN, suite models were created by merging 101,860 unstable cells (about 80% of the total area) and an equal number of stable cases. The negative cases in the subset were randomly selected in order to prevent stable cells from overlapping.

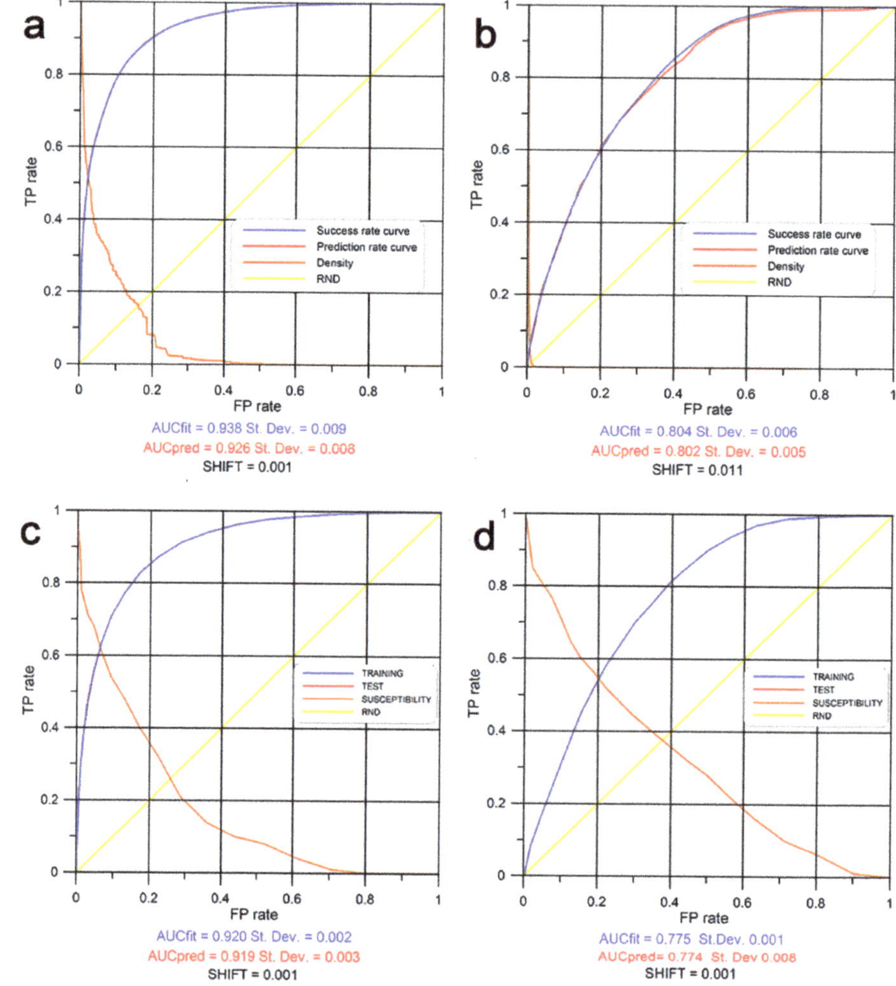

Figure 4. Receiver operating characteristic (ROC) curves: (**a**) conditional analysis (CA) for scarp landslide (SCR_LSN); (**b**) CA for hillslope landslide (HILL_LSN); (**c**) binary logistic regression (BLR) analysis for SCR_LSN; (**d**) BLR analysis for HILL_LSN.

Categorical variables were binarized and BLR was applied. Each model underwent the BLR procedure 10 times. An open source statistical package (TANAGRA) was used to generate the contingency tables (Table 4) and automatically extract the true positive (TP) and true negative (TN), and single estimates of the sensitivity or hit rate (TP/(TP + FN)) and 1-specificity (FP/(TP + FN); [66–70].

For SCR_LSN, drawing from 18 predictors, 13 were always selected in all 100 model repeats of the models. SLO, LCL_MNTPS, and USE_321 were systematically extracted as the first three significant factors, and with positive coefficients. The regression coefficients, for each selected predictor, were marked by conceptually coherent signs. A scan was easily imagined, and LCL_PLAINS and LCL_OPEN had a negative coefficient instead (Table 5).

Table 4. Contingency table for the 10-model suite. a for the SCR_LSN; b for HILL_SLN. TP = true positives, TN = true negatives, FP = false positives, FN = false negatives.

(a)		TEST SUBSET-PREDICTION SKILL									
		PREDICTED YES		PREDICTED NO		RECALL		FALL-OUT		ERROR RATE	AUC
		YES	NO	YES	NO	YES TP/oP	NO TN/oN	YES FP/pP	NO FN/pN		
MODELS	1	13,648	2850	3011	13,861	0.819	0.829	0.173	0.178	0.176	0.916
	2	13,672	2853	3008	13,837	0.820	0.829	0.173	0.179	0.176	0.916
	3	13,849	2686	2831	14,004	0.830	0.839	0.162	0.168	0.165	0.921
	4	13,802	2634	2878	14,056	0.827	0.842	0.160	0.170	0.165	0.923
	5	13,686	2814	2994	13,876	0.821	0.831	0.171	0.177	0.174	0.917
	6	13,735	2743	2945	13,947	0.823	0.836	0.166	0.174	0.170	0.919
	7	13,732	2817	2948	13,873	0.823	0.831	0.170	0.175	0.173	0.917
	8	13,794	2755	2896	13,935	0.826	0.835	0.166	0.172	0.169	0.918
	9	13,790	2622	2890	14,068	0.827	0.843	0.160	0.170	0.165	0.923
	10	13,781	2688	2902	14,002	0.826	0.839	0.163	0.172	0.168	0.917
ALL		13,748.9	2746.2	2930.3	13,945.9	0.824	0.835	0.166	0.174	0.170	0.919
		65.1	86.2	60.9	83.6	0.004	0.005	0.005	0.004	0.004	0.003

(b)		TEST SUBSET-PREDICTION SKILL									
		PREDICTED YES		PREDICTED NO		RECALL		FALL-OUT		ERROR RATE	AUC
		YES	NO	YES	NO	YES TP/oP	NO TN/oN	YES FP/pP	NO FN/pN		
MODELS	1	206,596	100,179	48,054	154,471	0.811	0.607	0.327	0.237	0.291	0.776
	2	206,596	100,179	48,054	154,471	0.811	0.607	0.327	0.237	0.291	0.775
	3	207,100	100,614	47,550	154,036	0.813	0.605	0.327	0.236	0.291	0.776
	4	206,960	101,894	47,550	154,036	0.813	0.602	0.330	0.236	0.293	0.772
	5	206,275	99,734	48,375	154,916	0.810	0.608	0.326	0.238	0.291	0.776
	6	206,665	100,608	47,659	153,889	0.813	0.605	0.327	0.236	0.291	0.766
	7	206,991	100,761	47,659	153,889	0.813	0.604	0.327	0.236	0.291	0.775
	8	206,455	99,808	48,195	154,842	0.811	0.608	0.326	0.237	0.291	0.775
	9	206,755	101,015	47,895	153,635	0.812	0.603	0.328	0.238	0.292	0.773
	10	206,880	101,576	47,770	153,074	0.812	0.601	0.329	0.238	0.293	0.775
ALL		206,727	100,636	47,876	154,125.9	0.811	0.604	0.327	0.237	0.292	0.774
		259.0	710.5	285.3	562.8	0.001	0.002	0.001	0.001	0.001	0.003

The main significant controlling factors for SCR_LSN in the study area which showed a slope angle for LCL_MNTPS was landform classification (positive coefficient), and LCL_PLAINS with a negative coefficient.

On to the HILL_LSN, the predictors extracted by BLR, are more than those extracted for SCR LSN, and above all, this slope angle variable does not appear to be as highly significant as may have been imagined. This is probably due to the fact that the diagnostic areas used here (whole landslide body area) are not so discriminant and able to determine the preparatory conditions for landslides. Additionally, it may owe something to the reasonably large cell size. The PAI landslide inventory is not characterized by high accuracy and some landslide typologies, like for "Areas with diffused landslide", do not allow a true discrimination of the geo-environmental variables which influence the slope stability conditions. Among the variables, LCL PLAINS ranks as the first predictor extracted in the analysis with a negative coefficient. Among all the variables, 20 were systematically extracted in all 100 different repeats, lithological conditions being those selected most frequently, and with a higher weight. Among them, LITH_Ca is characterized by having a negative coefficient value (Table 6).

Table 5. Factors analysis for SCR_LSN.

FACTORS ANALYSIS							
SCR_LSN							
Attribute	**Avg. (β)**	**Avg. (Std-dev)**	**Avg. Wald**	**Avg. Signif**	**Avg. OR**	**Avg. RK**	**n**
SLO	0.111	0.003	1046.925	0.0000	1.118	1.0	100
LCL_MNTPS	0.822	0.084	95.372	0.0000	2.281	2.0	100
USE_321	0.977	0.070	195.690	0.0000	2.664	3.0	100
USE_111	2.237	0.207	116.737	0.0000	9.419	4.7	100
LCL_PLAINS	−2.071	0.164	159.161	0.0000	0.127	5.1	100
LITH_Ca	0.624	0.091	53.558	0.0000	1.886	8.3	100
LITH_Ev	0.934	0.120	63.700	0.0000	2.596	8.5	100
USE_23	0.599	0.084	51.154	0.0000	1.827	9.4	100
USE_112	1.571	0.220	51.222	0.0000	4.901	9.7	100
LCL_OPEN	−0.562	0.072	61.954	0.0000	0.570	10.0	100
LITH_SaCl	−0.644	0.111	37.426	0.0062	0.531	10.1	100
LITH_PhMe	−0.898	0.143	42.396	0.0002	0.413	10.3	100
LCL_USHP	−0.743	0.178	18.055	0.0005	0.480	13.8	100
LITH_CDC	0.650	0.130	25.732	0.0003	1.947	12.0	97
USE_22	0.308	0.095	10.624	0.0024	1.362	15.2	52
LITH_Cl	−0.028	0.106	5.054	0.2675	1.008	9.3	38
LITH_CoSa	0.576	0.179	10.373	0.0030	1.802	16.3	25
LITH_SaCa	0.329	0.156	7.230	0.1161	1.427	13.3	14

Table 6. Factors analysis for HILL_LSN.

FACTORS ANALYSIS							
HILL_LSN							
Attribute	**Avg. (β)**	**Avg. (Std-dev)**	**Avg. Wald**	**Avg. Signif**	**Avg. OR**	**Avg. RK**	**n**
LCL_PLAINS	−1.7649	0.0561	5788.5652	0.0050	0.1807	1.0	100
RAIN_H	0.0042	0.0001	3906.3562	0.0000	1.0042	2.0	100
LITH_Cl	4.2016	0.0893	2217.0432	0.0000	66.9131	3.0	100
LITH_SaCl	4.0756	0.0898	2062.8770	0.0000	58.9854	4.1	100
LITH_CoSa	4.3585	0.0930	2200.4857	0.0000	78.2727	4.9	100
LITH_Ev	4.0339	0.0927	1895.9701	0.0000	56.5819	6.0	100
LCL_MNTPS	−0.5599	0.0548	775.7442	0.0334	0.6012	7.2	100
USE_22	−0.5557	0.0242	930.3611	0.0031	0.5768	8.0	100
USE_31	−0.7374	0.0296	930.1456	0.0000	0.4809	8.8	100
LITH_PhMe	3.6490	0.0927	1552.2501	0.0000	38.5212	10.4	100
LITH_CDC	3.6064	0.0919	1543.0805	0.0000	36.8982	11.4	100
LITH_Ca	−3.3053	0.0903	1343.2796	0.0000	27.3051	12.4	100
LITH_SaCa	3.2414	0.0923	1234.5050	0.0000	25.6089	13.4	100
LCL_UPPSL	−0.5569	0.0665	311.2864	0.0315	0.6040	14.4	100
USE_22	−0.6382	0.0565	133.4693	0.0000	1.9036	16.7	100
SLO	0.0130	0.0009	196.2342	0.0000	1.0130	16.8	100
LCL_USHP	0.4612	0.0612	171.5003	0.0001	1.6692	17.4	100
LCL_MRDG	−0.3298	0.0699	106.7273	0.0154	0.7555	18.7	100
USE_51	−1.2700	0.1325	94.3962	0.0000	0.2827	20.1	100
USE_32	−0.1319	0.0225	85.1833	0.0019	0.8815	20.7	100
USE_14	−11.0917	79.2952	0.0235	0.8809	0.0000	22.9	94
LCL_CANY	0.3067	0.0563	114.4982	0.0005	1.4338	19.4	90
USE_33	−0.4630	0.1721	7.4578	0.0137	0.6318	12.3	47
USE_112	0.2241	0.0632	11.5918	0.0035	1.2653	23.0	43
USE_13	−0.3685	0.1254	8.7318	0.0044	0.6923	23.6	31
USE_21	0.0654	0.0419	10.9851	0.0024	1.0881	23.9	24
LSC_OPEN	0.2898	0.1851	63.6125	0.0693	1.5858	17.4	20
LCL_MDRG	0.2982	0.2121	25.2307	0.0123	1.6133	21.1	18
USE_23	0.3639	0.0903	15.6449	0.0003	1.4481	23.9	8

6. Discussion

These analyses allow the generation of the susceptibility maps for SCR_LSN and HILL_LSN for both CA and for BLR.

The GRID UCU layers were intersected with 100 different TRNsubset landslide layers (SCR_LSN and HILL_SCR) and the mean density was calculated for each mapping unit. The susceptibility model derived from the CA approach was obtained by assigning each UCU to its corresponding class, according to its computed density. For each landslide type, the computed density corresponds to the susceptibility function (S_{UCU}) equivalent to the conditional probability of a new landslide, given the selected predictive variables [44].

Although numerous studies on landslide susceptibility zoning have been published, no global approach is yet shared by the scientific community to classify susceptibility maps. In the studies, where the classification of territories in accordance with their level of landslide susceptibility is the aim, it is appropriate to determine the optimal cutoff classification values which are capable of dividing the mapping units mainly into two large domains: stable areas (susceptibility values are less than the cutoff) to the left of the cutoff and unstable terrain (susceptibility greater than the cutoff) to the right of the cutoff [66]. When statistical approaches such as CA and BLR are used, the statistical software sets the significant cutoff (*m*cutoff) as equal to 0.5 by default [54,71–74] in order to correctly classify the predicted stable or unstable cells. This research area is quite extensive (>25,000 km^2), therefore, it is useful to divide the territory into a few different classes, depending on the value of susceptibility, and define some useful class range boundaries for the susceptibility (or density) index. The aim is to be able to divide the entire Sicilian territory into four classes of susceptibility: very low, moderate, high, and very high. To such an end, three cutoff values are required: the low cutoff (*l*cutoff), the central cutoff (*m*cutoff), and the high cutoff (*h*cutoff). Operationally, the *m*cutoff is first identified graphically, on the ROC curves, as the maximum value of the difference between the FP and TP rate. Subsequently, with the same method, were identified the other limits of the classes for the areas to the right of the *m*cuoff (high) and for the areas of the left of the *m*cuoff (low).

Table 7 shows the identification of *m*cutoff for the model obtained by following the CA for SCR_LSN. The susceptibility maps based on both statistical approaches have been derived and with test inventory subsets verified. Therefore, the maps presented in Figure 5 classify the Sicilian territory with low, moderate, high and very high susceptibility values according to their degree of propensity to instability and second cutoff probability values identified objectively.

Table 7. Cutoff range for (a) HILL_LSN and (b) SCR_LSN.

(a)		CA	
		Landslide Tipology	
		SCR_LSN	**HILL_LSN**
Classes	Very Low	0–0.34	0–0.04
	Moderate	0.34–2.5	0.04–0.07
	High	2.5–12	0.07–0.16
	Very High	0.78–1.00	0.16–1.00
(b)		**BLR**	
		Landslide Tipology	
		SCR_LSN	**HILL_LSN**
Classes	Very Low	0–0.10	0–0.10
	Moderate	0.1–0.15	0.1–0.28
	High	0.15–0.48	0.28–0.48
	Very High	0.48–1.00	0.48–1.00

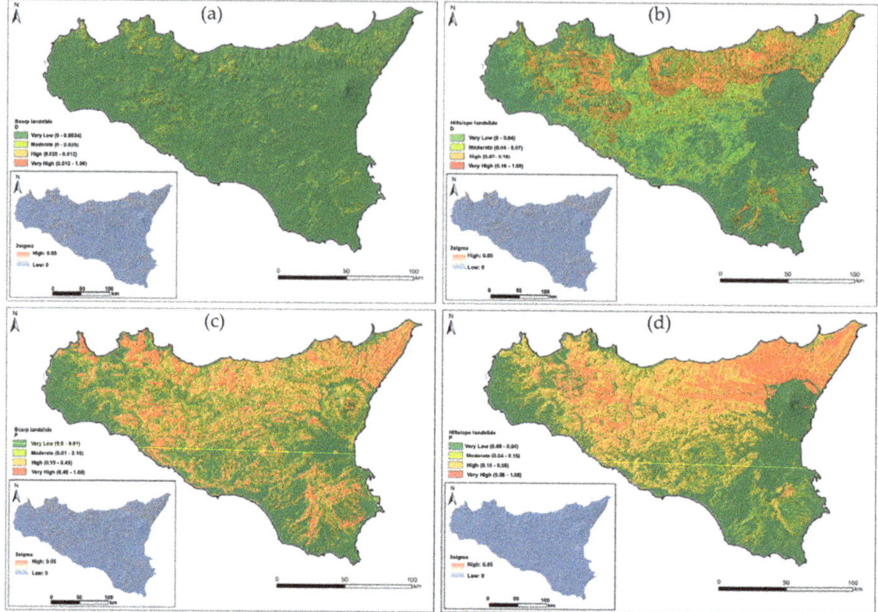

Figure 5. Susceptibility distribution for HILL_LSN using CA (**a**) and BLR analysis (**b**); susceptibility distribution for SCR_LSN using CA (**c**) and BLR analysis (**d**).

The presented research is focused on verifying two main concepts: the first, to explore the possibility of using existing data (landslide inventories, thematic maps of predictor variables) to create regional landslide susceptibility models; the second, to verify the exploitability of CA, comparing it with models based on BLR, in order to create landslide susceptibility models for areas and studies bigger than 1:100,000.

When analyzing the maps presented in Figure 5, it is noteworthy that those created by BLR (b, d) have a chromatic gradation indicating susceptibility conditions for Sicily which are generally higher than those created by CA (a, c). Therefore, this remains an open question: Which statistical approach is more representative of the conditions of susceptibility?

To answer this question, we analyzed the various sectors of the research area in detail and the different maps were compared and analyzed (Figure 6).

In order to single out the targets in the present study, 75% of instability phenomena (training subset) surveyed in the regional inventory of landslides, produced by the Environment and Territory Department of the Sicilian region (ARTA) was used to create two different landslide susceptibility maps: one based on the conditional analysis approach, and another on the binary logistic regression approach. The slope failure archive was simplified into two main types based on expert judgment of which preparatory variables they shared: the scarp landslide and hillslope landslide. The results obtained by CA showed an outstanding predictive ability for models based on a small number of predictive parameters combined in UCUs which were verified through spatial validation using a test subset randomly extracted from the Sicilian regional landslide inventory that covers the entire territory. ROC curve validation of the 100 different models showed the unquestionable excellence and stability of the forecasting performance for both SCR_LSN and HILL_LSN. A quality control test on four susceptibility maps was applied according to the degree of fit approach (DF) [24,27,73–75]. DF represents the percentage of an area subject to landslides for each range of classes of susceptibility and is determined by cross-tabulation of the landslide test subset with the susceptibility class of maps. DF can be expressed with the following formula:

$$DF = \frac{LSNi/Si}{\sum LSNi/Si} \quad (3)$$

where LSN is the area occupied in the i-class of susceptibility and Si is the area of the i susceptibility class [74,75].

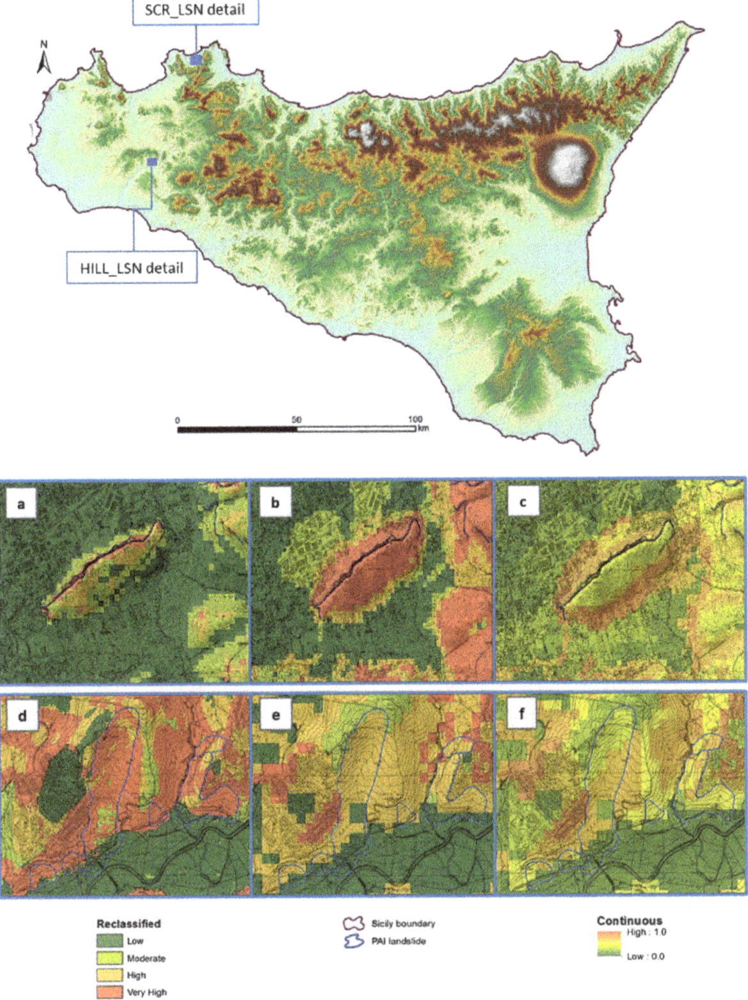

Figure 6. Different classifications of susceptibility values for HILL_LSN (**d**,**e**,**f**) and SCR_LSN (**a**,**b**,**c**) using BLR (**a**,**d**) and CA (**b**,**e**) reclassified. Figures (**c**) and (**f**) show the susceptibility values obtained with the BLR technique and are shown with continuous values.

The percentage of area subject to landslides falling into the null or moderate susceptibility class may be considered as a false negative error. The histograms shown in Figure 7 highlight an excellent predictive ability of new landslides not seen in the construction of maps. The percentages of relative accuracy are understood as the sum of the degrees of fit of the high and very high susceptibility classes [27] and the value of R^2 of the trend line confirms the goodness of the susceptibility performance models created.

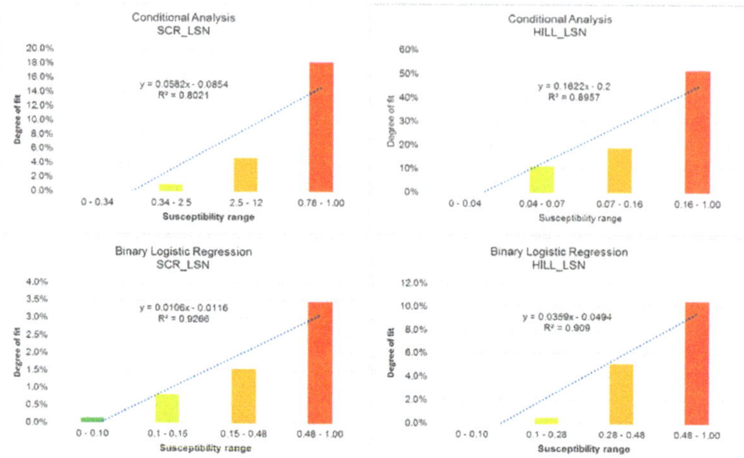

Figure 7. Distribution of landslides by susceptibility classes.

The propensity of a territory to be affected by new landslides and the degree of hazard or risk that characterizes it are usually expressed with the help of a map in which the area is divided into different zones according to the different values that qualify it. In this mapping, the territory is zoned or divided into homogeneous zones or user-defined fields/areas, the ranking of which is defined according to their real or potential degree of landslide susceptibility [8].

From among the Italian regions, Sicily is one of areas most affected by geomorphological instability. Landslide activity is a clear threat to the territory, facilities, and people present there. From this point of view, each part of the territory is characterized by a landslide vulnerability value. Generally, it depends on the territorial level of exposure to the threat, determined by the socio-economic value of the assets as well as by their resistance to the stresses expected. Interaction between humans and the natural environment is a very complex and diverse issue, not often approached in a systematic way, as resources are primarily invested in risk-mitigating measures, while being severely limited when it comes to the medium- and long-term research needed to understand the environment better and more effectively. The current PAI archive version is highly dependent on the past instability scenario. By looking at those that were counted and catalogued using a matrix system of evaluation, it is possible to derive the conditions of the associated geomorphological risk. This last concept represents a big step forward as it is now necessary to consider strongly the concept of landslide susceptibility which would represent the adoption of a spatial analysis tool with predictive power. On this scale, output cuts will be based on administrative boundaries (municipalities).

To meet this goal, the mean values of susceptibility (CA) and probability (BLR) were assigned to each municipality of the Sicily region (the calculation was only performed for the 382 municipalities included within the main island) in a GIS environment.

The output susceptibility values obtained by means of the two methods (CA and BLR) for the municipal boundaries are represented as maps in Figure 8. The picture shows the susceptibility maps relating to both conditional methods (a, b) and BLR (c, d) for the SCR_LSN and HILL_LSN typologies, respectively. In particular, it is possible to observe that for the SCR_LSN types, maps (a, c) are very different with regard to the distribution of the susceptibility values among the municipal boundaries. On the other hand, the comparison between the maps concerning the HILL_LNS type shows a convergence among the results and suitable differentiation for the urban units. Both statistical approaches (CA and BLR) show higher susceptibility landslide values in the northeastern portion of the island, in correspondence with the mountainous chains of Nebrodi and Peloritani in the province of Messina.

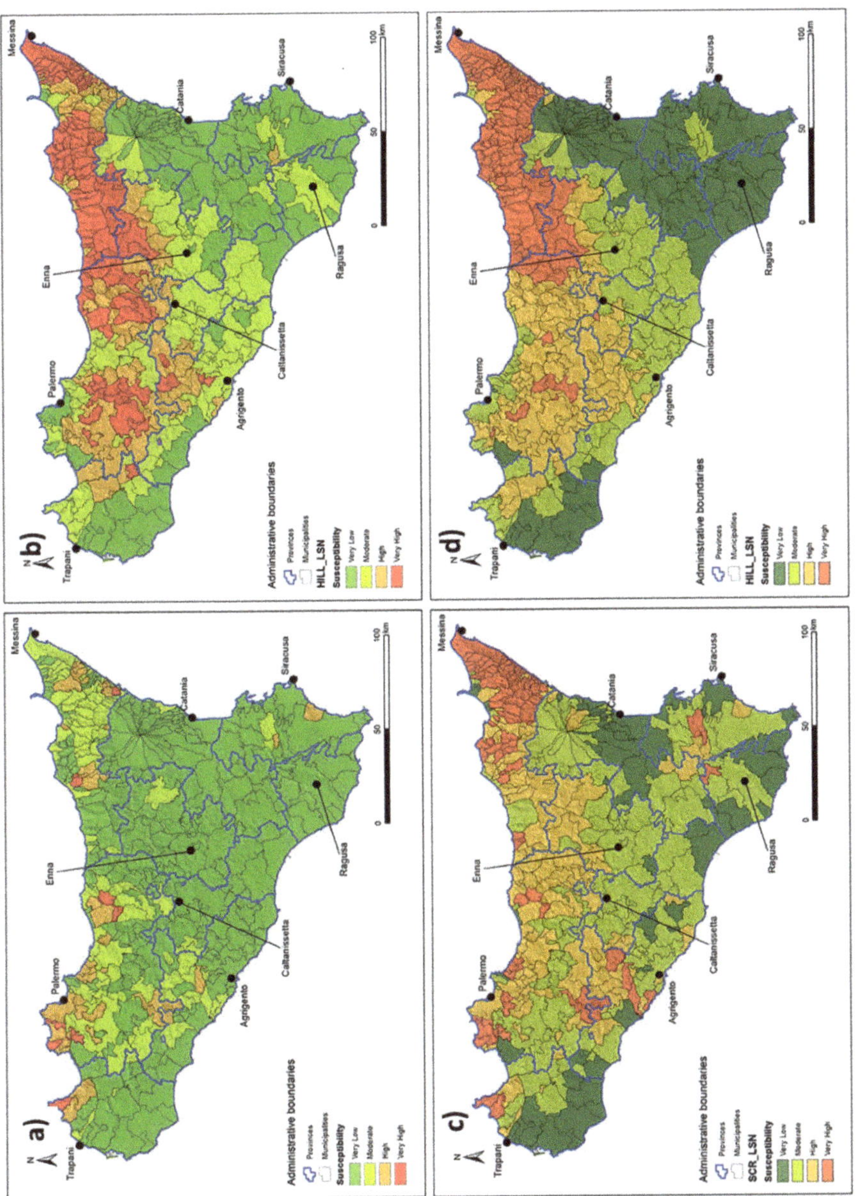

Figure 8. Regional susceptibility map with municipalities boundary. **(a,b)** Using the CA approach for SCR_LSN; **(c,d)** showing the susceptibility classification using the BLR approach.

7. Final Remarks and Management Implications

The research results presented here have demonstrated the ability to leverage a set of existing data already in the public administration to generate regional landslide susceptibility maps. A total of 100 performances executed for both the CA and BLR approaches have allowed susceptibility patterns characterized by excellent performance to be obtained, with remarkable stability, robustness, and reliability, even if CA generated more powerful models in terms of forecasted performance (AUC curves). With the goal of creating, for the whole Sicilian territory, susceptibility maps that could be useful for the purposes of effective land management and spatial urban planning, this study proposes a reclassification method into four susceptibility map classes. The reclassified susceptibility maps, determined by the identified cutoff boundaries, were validated quantitatively with the degree of fit technique. Validations demonstrated the reliability of the maps created and the explorability of the proposed protocol. However, some differences may be highlighted when comparing the final maps generated with two different approaches. In fact, the maps generated with CA seem to better represent the conditions leading to the failure of a territory than those created by BLR-generated ones. This is more realistic for the SCR_LSN than for HILL_LSN. This is probably due to the fact that the BLR tries to find a correlation equation between landslides and predictor training models, using a diagnostic area (the entire area subject to landslides) not only representative of the trigger conditions, involving areas only passively affected by the landslide and, in some cases generating an overestimation of the susceptibility value.

Of the 382 of Sicilian municipalities examined, 244 (nearly 64%) were correctly classified as unstable (i.e., prone to landslide instability) as they fall in the classes of high and/or very high susceptibility, both for scarp and for hillslope landslides. We particularly want to highlight that susceptibility is predicted as being very high in 90 of 108 municipalities in the province of Messina (more than 80%). The municipalities featuring lower susceptibility values were those within the provinces of Syracuse and Ragusa. This approach is based on the propensity for gravitational instability, unlike PAI, and allows the concept of landslide density or density index to be overcome. Table 8 lists the top 10 municipalities in order, compared to the size of an area that may be affected by new landslides in the future.

For each municipality surface area, the extension area has been reported, the density (D) which corresponds to the landslide density for the current phenomena surveyed in the PAI and the P (or the area which is located to the right of *m*cutoff) for each municipality, and finally, the difference between the current framework of the landslide and that provided (Odd).

The table shows that the towns of Isola delle Femmine and Roccafiorita, which despite having a higher SCR_LSN density (respectively 7.04% and 6.04%) are the last two municipalities in the table, because of the difference, in terms of Odd, between the density and the future probability, which is lower than that seen in the municipalities of San Vito Lo Capo and Frazzanò. Almost 4 km of new territory in the municipality of San Vito Lo Capo, for example, could be affected by new SCR_LSN activations in the future. Similarly, we can say that linked to HILL_LSN, the territory of Alcara Li Fusi is the Sicilian municipality with the highest HILL_LSN density value, but the municipalities of Ficarra (with 8 km^2) and Sinagra (with 6.54 km^2), both in the province of Messina, will be most affected by new activations of hillslope landslides.

The large and widespread use of known geostatistical methods has gone through at least three decades of landslide hazard studies, but still does not eliminate some of the conceptual and operational bottlenecks, only sporadically resulting in the safety enforcement schemes by the authorities involved in studying landslide risk in Italy.

Table 8. The top 10 municipalities ordered according to the size of the area that may be affected by new landslides in the future ((a) SCR_LSN in Table; (b) HILL_LSN).

(a) Munipality	Area (km^2)	P	D	odd	km^2
SAN VITO LO CAPO	3.6	4.42%	1.98%	2.44%	3.91
FRAZZANO'	1.2	5.90%	3.02%	2.88%	1.46
GIARDINELLO	50.8	5.76%	2.33%	3.42%	0.89
ISNELLO	6.9	6.76%	3.70%	3.05%	0.82
TORRETTA	26.0	4.28%	1.61%	2.67%	0.69
SAN MARCO D'ALUNZIO	12.8	5.85%	2.58%	3.27%	0.23
BORGETTO	26.0	5.12%	2.97%	2.16%	0.14
CINISI	33.0	4.81%	2.57%	2.24%	0.14
ISOLA DELLE FEMMINE	60.2	10.70%	7.04%	3.66%	0.13
ROCCAFIORITA	25.5	9.49%	6.04%	3.45%	0.04
(b) Munipality	**Area (km^2)**	**P**	**D**	**odd**	**km^2**
FICARRA	215.6	88.85%	3.72%	85.13%	8.03
SINAGRA	70.0	93.26%	9.34%	83.91%	6.54
ALCARA LI FUSI	31.1	91.98%	16.74%	75.24%	5.21
CASTELL'UMBERTO	30.2	92.05%	7.31%	84.74%	2.21
SAN PIERO PATTI	14.4	89.30%	9.77%	79.52%	1.40
MONTAGNAREALE	11.4	89.73%	11.07%	78.66%	1.26
RACCUJA	15.8	88.78%	5.60%	83.19%	0.88
UCRIA	26.1	94.39%	3.14%	91.25%	0.82
SANT'ANGELO DI BROLO	18.5	89.90%	4.34%	85.56%	0.80
TORTORICI	23.9	89.21%	2.76%	86.46%	0.66

Since economic problems, which are common to all countries, do not allow either investment in research projects on a medium- and long-term scale, the concept of landslide susceptibility should represent, for all political and administrative actors dealing with environmental and territorial policies, a new approach to the problem associated with mass movements. For this reason, the scientific community is engaged in a continuous search for methods and techniques to estimate the degree of real and potential instability, using the minimum amount of equipment, data, and economic resources possible. Generally, substantial difficulty exists in identifying the most reliable procedures, allowing this matter to be approached in a non-traditional manner based on modelling and investigative techniques built on the exchange of experiences between experts, studies, and experiments on every continent, and showing different strategies and possible technical combinations, depending on the type and/or the number and complexity of the investigation, producing susceptibility, hazard, and risk maps, used as the basis for decision-making processes in land management. In this framework, further effort is needed in trying to make the different methods more objective and shared by all, in order to be simple and repeatable and, most of all, in transferring the knowledge gained to laws that underpin territorial planning, building regulations, and in civil defense plans [22].

Over the decades, many research groups and national and international commissions have tried to provide precise definitions, in order to generate maps indicating the different urban planning vocation of an area. For this reason, the scientific community is engaged in a continuous search for methods and techniques to estimate the degree of real and potential instability, using the minimum amount of equipment and possible economic resources.

Author Contributions: Writing–original draft, D.C.; review & editing, C.I. All authors have read and agreed to the published version of the manuscript.

Funding: This research received no external funding.

Conflicts of Interest: The authors declare no conflict of interest.

References

1. Guzzetti, F.; Carrara, A.; Cardinali, M.; Reichenbach, P. Landslide hazard evaluation: A review of current techniques and their application in a multi-scale study, Central Italy. *Geomorphology* **1999**, *31*, 181–216. [CrossRef]
2. Guzzetti, F.; Reichenbach, P.; Cardinali, M.; Galli, M.; Ardizzone, F. Probabilistic landslide hazard assessment at the basin scale. *Geomorphology* **2005**, *72*, 272–299. [CrossRef]
3. Brabb, E.E. Innovative approach to landslide hazard and risk mapping. In Proceedings of the 4th International Symposium on Landslides, Toronto, ON, Canada, 16–21 September 1984; Volume 1, pp. 307–324.
4. Soeters, R.; van Westen, C.J. Slope instability recognition, analysis, and zonation. In *Landslides Investigation and Mitigation*; Special Report 247; Transportation Research Board: Washington, DC, USA, 1996; pp. 129–177.
5. Lyell, C. *Principles of Geology, Being an Attempt to Explain the Former Changes of the Earth's Surface, by Reference to Causes Now in Operation*; John Murray: London, UK, 1833; Volume 3.
6. Aleotti, P.; Chowdhury, R. Landslide hazard assessment: Summary review and new perspectives. *Bull. Eng. Geol. Environ.* **1999**, *58*, 21–44. [CrossRef]
7. Carrara, A.; Cardinali, M.; Detti, R.; Guzzetti, F.; Reichenbach, P. Gis techniques and statistical models in evaluating landslide-hazard. *Earth Surf. Process. Landf.* **1991**, *16*, 427–445. [CrossRef]
8. Varnes, D.J. Slope movements, type and processes. In *Landslides Analysis and Control*; Schuster, R.L., Krizek, R.J., Eds.; Special Report 176; Transportation Research Board: Washington, DC, USA, 1984.
9. Cascini, L. Applicability of landslide susceptibility and hazard zoning at different scales. *Eng. Geol.* **2008**, *102*, 164–177. [CrossRef]
10. Hervás, J.; Günther, A.; Reichenbach, P.; Chacón, J.; Pasuto, A.; Malet, J.-P.; Trigila, A.; Hobbs, P.; Maquaire, O.; Tagliavini, F.; et al. Recommendations on a common approach for mapping areas at risk of landslides in Europe. In *Guidelines for Mapping Areas at Risk of Landslides in Europe*; Hervás, J., Ed.; JRC Report EUR 23093 EN; Office for Official Publications of the European Communities: Luxembourg, 2007; 53p.
11. Eckelmann, W.; Baritz, R.; Bialousz, S.; Carre, F.; Jones, B.; Kibblewhite, M.; Kozak, J.; Le Bas, C.; Tóth, G.; Várallyay, G.; et al. *Common Criteria for Risk Area Identification according to Soil Threats*; European Soil Bureau Research Report No.20, EUR 22185 EN; Office for Official Publications of the European Communities: Luxembourg, 2006; 94p.
12. Glade, T.; Reichenbach, P.; Resources, N. Preface "Landslide hazard and risk assessment at different scales". *Nat. Hazards Earth Syst. Sci.* **2013**, *13*, 2169–2171.
13. Günther, A.; Reichenbach, P.; Malet, J.; Van Den Eeckhaut, M.; Hervás, J.; Dashwood, C.; Guzzetti, F. Tier-based approaches for landslide susceptibility assessment in Europe. *Landslides* **2012**, *10*, 529–546. [CrossRef]
14. Manzo, G.; Tofani, V.; Segoni, S.; Battistini, A.; Catani, F. GIS techniques for regional-scale landslide susceptibility assessment: The Sicily (Italy) case study. *Int. J. Geogr. Inf. Sci.* **2013**, *27*, 1433–1452. [CrossRef]
15. ARTA_Sicilia. *Piano Stralcio di bacino per l'Assetto Idrogeologico Della Regione Siciliana*; Assessorato Regionale Territorio e Ambiente della Regione Sicilia: Sicilia, Italy, 2004; 165p.
16. Agnesi, V.; Di Maggio, C.; Macaluso, T. Morphostructural setting and geomorphological evolution in the Madonie Mountains (Northern Sicily, Italy). *Suppl. Geogr. Fis. Din. Quat.* **1997**, *3*, 43–44.
17. Catalano, R.; Agate, M.; Albanese, C.; Avellone, G.; Basilone, L.; Morticelli, G.; Gugliotta, C.; Sulli, A.; Valenti, V.; Gibilaro, C.; et al. *Walking along a Crustal Profile Across the Sicily Fold and Thrust Belt*; Geological Field Trips: Palermo, IT, USA, 2013; Volume 5.
18. Di Maggio, C.; Madonia, G.; Parise, M.; Vattano, M. Karst of Sicily and its conservation. *J. Cave Karst Stud.* **2012**, *74*, 157–172. [CrossRef]
19. Dikau, R.; Brunsden, D.; Schrott, L.; Ibsen, M. *Landslide Recognition. Identification, Movement and Courses*; Report No. 1 of the European Commission Environment Programme; Wiley and Sons: Chichester, UK, 1996; 251p.
20. Cruden, D.M. A simple definition of a landslide. *Bull. Int. Assoc. Eng. Geol.* **1991**, *43*, 27–29. [CrossRef]
21. Cruden, D.M.; Varnes, D.J. Landslides Types and Processes. In *Landslides: Investigation and Mitigation*; Special Report 247; Transportation Research Board: Washington, DC, USA, 1996; pp. 36–75.
22. Guzzetti, F.; Reichenbach, P.; Ardizzone, F.; Cardinali, M.; Galli, M. Estimating the quality of landslide susceptibility models. *Geomorphology* **2006**, *81*, 166–184. [CrossRef]

23. Bosi, C.; Dramis, F.; Gentili, B. Carte geomorfologiche di dettaglio e carte di stabilita: Esempi nel territorio marchigiano. *Geol. Appl. Idrogeol.* **1982**, *20*, 53–62.
24. Chacón, J.; Irigaray, C.; Fernández, T.; El Hamdouni, R. Engineering geology maps: Landslides and geographical information systems. *Bull. Eng. Geol. Environ.* **2006**, *65*, 341–411. [CrossRef]
25. Carrara, A.; Guzzetti, F.; Cardinali, M.; Reichenbach, P. Use of GIS technology in the prediction and monitoring of landslide hazard. *Nat. Hazards* **1999**, *20*, 117–135. [CrossRef]
26. Carrara, A.; Cardinali, M.; Guzzetti, F.; Reichenbach, P. GIS technology in mapping landslide hazard. In *Geographical Information Systems in Assessing Natural Hazards*; Carrara, A., Guzzetti, F., Eds.; Kluwer Academic Publishers: Dordrecht, The Netherlands, 1995; pp. 135–175.
27. Irigaray, C.; Fernández, T.; El Hamdouni, R.; Chacón, J. Evaluation and validation of landslide-susceptibility maps obtained by a GIS matrix method: Examples from the Betic Cordillera (southern Spain). *Nat. Hazards* **2007**, *41*, 61–79. [CrossRef]
28. Carrara, A.; Crosta, G.; Frattini, P. Comparing models of debris-flow susceptibility in the alpine environment. *Geomorphology* **2008**, *94*, 353–378. [CrossRef]
29. Lombardo, L.; Cama, M.; Maerker, M.; Rotigliano, E. A test of transferability for landslides susceptibility models under extreme climatic events: Application to the Messina 2009 disaster. *Nat. Hazards* **2014**, *74*, 1951–1989. [CrossRef]
30. Rossi, M.; Guzzetti, F.; Reichenbach, P.; Mondini, A.C.; Peruccacci, S. Optimal landslide susceptibility zonation based on multiple forecasts. *Geomorphology* **2010**, *114*, 129–142. [CrossRef]
31. Van Den Eeckhaut, M.; Hervás, J.; Jaedicke, C.; Malet, J.P.; Montanarella, L.; Nadim, F. Statistical modelling of Europe-wide landslide susceptibility using limited landslide inventory data. *Landslides* **2011**, *9*, 357–369. [CrossRef]
32. Clerici, A.; Perego, S.; Tellini, C.; Vescovi, P. Landslide failure and runout susceptibility in the upper T. Ceno valley (Northern Apennines, Italy). *Nat. Hazards* **2009**, *52*, 1–29. [CrossRef]
33. Conoscenti, C.; Di Maggio, C.; Rotigliano, E. GIS analysis to assess landslide susceptibility in a fluvial basin of NW Sicily (Italy). *Geomorphology* **2008**, *94*, 325–339. [CrossRef]
34. Costanzo, D.; Cappadonia, C.; Conoscenti, C.; Rotigliano, E. Exporting a Google EarthTM aided earth-flow susceptibility model: A test in central Sicily. *Nat. Hazards* **2011**, *61*, 103–114. [CrossRef]
35. Costanzo, D.; Rotigliano, E.; Irigaray, C.; Jiménez-Perálvarez, J.D.; Chacón, J. Factors selection in landslide susceptibility modelling on large scale following the gis matrix method: Application to the river Beiro basin (Spain). *Nat. Hazards Earth Syst. Sci.* **2012**, *12*, 327–340. [CrossRef]
36. Costanzo, D.; Chacón, J.; Conoscenti, C.; Irigaray, C.; Rotigliano, E. Forward logistic regression for earth-flow landslide susceptibility assessment in the Platani river basin (southern Sicily, Italy). *Landslides* **2014**, *11*, 639–653. [CrossRef]
37. Lombardo, L.; Cama, M.; Conoscenti, C.; Marker, M.; Rotigliano, E. Binary logistic regression versus stochastic gradient boosted decision trees in assessing landslide susceptibility for multiple-occurring landslide events: Application to the 2009 storm event in Messina (Sicily, southern Italy). *Nat. Hazards* **2015**, *79*, 1621–1648. [CrossRef]
38. Rotigliano, E.; Agnesi, V.; Cappadonia, C.; Conoscenti, C. The role of the diagnostic areas in the assessment of landslide susceptibility models: A test in the sicilian chain. *Nat. Hazards* **2011**, *58*, 981–999. [CrossRef]
39. Rotigliano, E.; Cappadonia, C.; Conoscenti, C.; Costanzo, D.; Agnesi, V. Slope units-based flow susceptibility model: Using validation tests to select controlling factors. *Nat. Hazards* **2012**, *61*, 143–153. [CrossRef]
40. Clerici, A.; Perego, S.; Tellini, C.; Vescovi, P. A GIS-based automated procedure for landslide susceptibility mapping by the Conditional Analysis method: The Baganza valley case study (Italian Northern Apennines). *Environ. Geol.* **2006**, *50*, 941–961. [CrossRef]
41. Vergari, F.; Della Seta, M.; Del Monte, M.; Fredi, P.; Lupia Palmieri, E. Landslide susceptibility assessment in the Upper Orcia Valley. *Nat. Hazards Earth Syst. Sci.* **2011**, *11*, 1475–1495. [CrossRef]
42. Davis, J.C. *Statistics and Data Analysis in Geology*; John Wiley & Sons Inc.: New York, NY, USA, 1973; 564p.
43. Chung, C.F.; Fabbri, A.G. Validation of Spatial Prediction Models for Landslide Hazard Mapping. *Nat. Hazards* **2003**, *30*, 451–472. [CrossRef]
44. Jiménez-Perálvarez, J.D.; Irigaray, C.; El Hamdouni, R.; Chacón, J. Landslide-susceptibility mapping in a semi-arid mountain environment: An example from the southern slopes of Sierra Nevada (Granada, Spain). *Bull. Eng. Geol. Environ.* **2010**, *70*, 265–277. [CrossRef]

45. Santacana, N.; Baeza, B.; Corominas, J.; De Paz, A.; Marturiá, J. A GIS-Based Multivariate Statistical Analysis for Shallow Landslide Susceptibility Mapping in La Pobla de Lillet Area (Eastern Pyrenees, Spain). *Nat. Hazards* **2003**, *30*, 281–295. [CrossRef]
46. Hosmer, D.W.; Lemeshow, S. Interpretation of the fitted logistic regression model. In *Applied Logistic Regression*, 2nd ed.; John Wiley & Sons, Inc.: New York, NY, USA, 2000; pp. 47–90.
47. Akgün, A.; Türk, N. Mapping erosion susceptibility by a multivariate statistical method: A case study from the Ayvalik region, NW Turkey. *Comput. Geosci.* **2011**, *37*, 1515–1524. [CrossRef]
48. Ayalew, L.; Yamagishi, H. The application of GIS-based logistic regression for landslide susceptibility mapping in the Kakuda-Yahiko Mountains, Central Japan. *Geomorphology* **2005**, *65*, 15–31. [CrossRef]
49. Kavzoglu, T.; Sahin, E.K.; Colkesen, I. Landslide susceptibility mapping using GIS-based multi-criteria decision analysis, support vector machines, and logistic regression. *Landslides* **2013**, *11*, 425–439. [CrossRef]
50. Van Den Eeckhaut, M.; Hervás, J. State of the art of national landslide databases in Europe and their potential for assessing landslide susceptibility, hazard and risk. *Geomorphology* **2012**, *139–140*, 545–558. [CrossRef]
51. Dai, F.C.; Lee, C.F.; Li, J.; Xu, Z.W. Assessment of landslide susceptibility on the natural terrain of Lantau Island, Hong Kong. *Environ. Geol.* **2001**, *40*, 381–391.
52. Kim, G.; Lee, J.; Lee, K. Application of representative elementary area (REA) to lineament density analysis for groundwater implications. *Water Resour.* **2004**, *8*, 27–42. [CrossRef]
53. Dai, F.C.; Lee, C.F. Landslide characteristics and slope instability modeling using GIS, Lantau Island, Hong Kong. *Geomorphology* **2020**, *42*, 213–228. [CrossRef]
54. Rakotomalala, R. TANAGRA: A free software for research and academic purposes. In Proceedings of the European Grid Conference, Amsterdam RNTI-E-3, Amsterdam, The Netherlands, 5 February 2005; Volume 2, pp. 697–702.
55. Crozier, M.J. Field assessment of slope instability. In *Slope Instability*; Brunsden, D., Prior, D.B., Eds.; John Wiley & Sons: Chichester, UK, 1984; pp. 103–142.
56. Hansen, A. Landslide hazard analysis. In *Slope Instability*; Brunsden, D., Prior, D.B., Eds.; John Wiley & Sons: Chichester, UK, 1984; pp. 523–602.
57. Palenzuela, J.A.; Jiménez-Perálvarez, J.D.; Chacón, J.; Irigaray, C. Assessing critical rainfall thresholds for landslide triggering by generating additional information from a reduced database: An approach with examples from the Betic Cordillera (Spain). *Nat. Hazards* **2016**, *84*, 185–212. [CrossRef]
58. Jenness, J. Topographic Position Index (TPI) v. 1.3a. Jenness Enterprises. 2006. Available online: http://www.jennessent.com/arcview/tpi.htm (accessed on 8 June 2020).
59. Regmi, N.R.; Giardino, J.R.; McDonald, E.V.; Vitek, J.D. A comparison of logistic regression-based models of susceptibility to landslides in western Colorado, USA. *Landslides* **2013**, *11*, 247–262. [CrossRef]
60. Hanley, A.J.; McNeil, J.B. The Meaning and Use of the Area under a Receiver Operating Characteristic (ROC) Curve. *Radiology* **1982**, *143*, 29–36. [CrossRef] [PubMed]
61. Fawcett, T. An introduction to ROC analysis. *Pattern Recognit. Lett.* **2006**, *27*, 861–874. [CrossRef]
62. Nandi, A.; Shakoor, A. A GIS-based landslide susceptibility evaluation using bivariate and multivariate statistical analyses. *Eng. Geol.* **2010**, *110*, 11–20. [CrossRef]
63. Frattini, P.; Crosta, G.; Carrara, A. Techniques for evaluating the performance of landslide susceptibility models. *Eng. Geol.* **2010**, *111*, 62–72. [CrossRef]
64. Lasko, T.A.; Bhagwat, J.G.; Zou, K.H.; Ohno-Machado, L. The use of receiver operating characteristic curves in biomedical informatics. *J. Biomed. Inform.* **2005**, *38*, 404–415. [CrossRef] [PubMed]
65. Goodenough, D.J.; Rossmann, K.; Lusted, L.B. Radiographic applications of receiver operating characteristic (ROC) curves. *Radiology* **1974**, *110*, 89–95. [CrossRef]
66. Egan, J.P. *Signal Detection Theory and ROC Analysis*; Academic Press: New York, NY, USA, 1975; 277p.
67. Williams, C.J.; Lee, S.S.; Fisher, R.A.; Dickerman, L.H. A comparison of statistical methods for prenatal screening for Down syndrome. *Appl. Stoch. Models Data Anal.* **1999**, *15*, 89–101. [CrossRef]
68. Nefeslioglu, H.A.; Gokceoglu, C.; Sonmez, H. An assessment on the use of logistic regression and artificial neural networks with different sampling strategies for the preparation of landslide susceptibility maps. *Eng. Geol.* **2008**, *97*, 171–191. [CrossRef]
69. Ohlmacher, G.C.; Davis, J.C. Using multiple logistic regression and GIS technology to predict landslide hazard in northeast Kansas, USA. *Eng. Geol.* **2003**, *69*, 331–343. [CrossRef]

70. Chung, C.F.; Fabbri, A.G.; Van Westen, C.J. Multivariate regression analysis for landslide hazard zonation. In *Geographical Information Systems in Assessing Natural Hazards*; Carrara, A., Guzzetti, F., Eds.; Kluwer Academic Publishers: Dordrecht, The Netherlands, 1995; pp. 107–133.
71. Atckinson, P.M.; Massari, R. Generalised linear modelling of susceptibility lo landsliding in the Central Apennines, Italy. *Comput. Geosci.* **1998**, *24*, 373–385. [CrossRef]
72. Irigaray, C.; Fernández, T.; El Hamdouni, R.; Chacón, J. Verification of landslide susceptibility mapping: A case study. *Earth Surf. Process. Landf.* **1999**, *24*, 537–544.
73. Fernández, T.; Irigaray, C.; El Hamdouni, R.; Chacon, J. Methodology for landslide susceptibility mapping by means of a GIS. Application to the Contraviesa area (Granada, Spain). *Nat. Hazard* **2003**, *30*, 297–308.
74. Remondo, J.; Gonzalez, A.; Diaz De Teran, J.R.; Cendrero, A.; Fabbri, A.; Chung, C.F. Validation of landslide susceptibility maps; examples and applications from a case study in Northern Spain. *Nat. Hazards* **2003**, *30*, 437–449. [CrossRef]
75. Plattner, T. Modelling public risk evaluation of natural hazards: A conceptual approach. *Nat. Hazards Earth Syst. Sci.* **2005**, *5*, 357–366. [CrossRef]

Article

Estimation of Actual Evapotranspiration Using the Remote Sensing Method and SEBAL Algorithm: A Case Study in Ein Khosh Plain, Iran

Amir Ghaderi [1,2], Mehdi Dasineh [3], Maryam Shokri [4] and John Abraham [5,*]

1 Department of Civil Engineering, Faculty of Engineering, University of Zanjan, Zanjan 4537138791, Iran; amir_ghaderi@znu.ac.ir
2 Department of Civil Engineering, University of Calabria, 87036 Arcavacata, Rende, Italy
3 Department of Civil Engineering, Faculty of Engineering, University of Maragheh, Maragheh 8311155181, East Azerbaijan, Iran; mehdi.dasineh3180@gmail.com
4 Architecture Department, School of Engineering, University College of Nabi Akram, Tabriz 513851488, Iran; shokri.maryam1991@gmail.com
5 School of Engineering, University of St. Thomas, St. Paul, MN 55105, USA
* Correspondence: jpabraham@stthomas.edu; Tel.: +1-612-963-2169

Received: 21 May 2020; Accepted: 22 June 2020; Published: 25 June 2020

Abstract: The aim of this study was to estimate evapotranspiration (ET) using remote sensing and the Surface Energy Balance Algorithm for Land (SEBAL) in the Ilam province, Iran. Landsat 8 satellite images were used to calculate ET during the cultivation and harvesting of wheat crops. The evaluation using SEBAL, along with the FAO-Penman–Monteith method, showed that SEBAL has a sufficient accuracy for estimating ET. The values of the Root Mean Square Error (RMSE), Mean Absolute Percentage Error (MAPE), Mean Bias Error (MBE), and correlation coefficient were 0.466, 2.9%, 0.222 mm/day, and 0.97, respectively. Satellite images showed that rainfall, except for the last month of cultivation, provided the necessary water requirements and there was no requirement for the use of other water resources for irrigation, with the exception of late May and early June. The maximum ET on the Ein Khosh Plain occurred in March. The irrigation requirements showed that the Ein Khosh Plain in March, which witnessed the highest ET, did not experience any deficiency of rainfall that month. However, during April and May, with maxima of 50 and 70 mm, respectively, water was needed for irrigation. During the plant growth periods, the greatest and least amount of water required were 231.23 and 19.47 mm/hr, respectively.

Keywords: Evapotranspiration; water requirement; remote sensing; SEBAL; Landsat 8

1. Introduction

Agriculture is one of the largest draws of freshwater resources. Because of the limited water availability, agricultural sectors have been forced to increase their efficiency. One way to improve water use management and increase the efficiency is to estimate the amount of water consumed by plants and the amount involved in evapotranspiration (*ET*) [1]. Knowledge of *ET* is important for modeling hydrologic fluxes and for proper water resource management. Spatial and temporal information on *ET* not only quantifies water loss caused by evaporation, but also provides information on the relationship between land use, water allocation, and water use [2].

In addition, an optimal use of water resources will save water during times when irrigation, for instance, is not needed. Reducing the use of water in times when it is not required will not only preserve water, but will also lower the soil water content and pore pressure. This, in turn, will improve the stability of the soil and make it more resistant to landslides or other soil instabilities.

In most parts of the world, post-rainfall *ET* is the second most important element of the water cycle, so an accurate estimation at the regional scale is essential to developing appropriate management strategies [3]. Due to the limited number of meteorological stations and the high cost of data collection, satellite data is often used to provide near real-time information on meteorological and environmental parameters. An important advantage of using remote sensing to aid water consumption is that *ET* can be measured without the need to quantify other complex hydrological processes [4]. The Surface Energy Balance Algorithm for Land (SEBAL) algorithm is one remote sensing algorithm that calculates the actual *ET* based on an instantaneous energy balance at the surface of each pixel from a satellite image. This technique has been applied in many countries with success [2]. Until now, different methods and sensors have been used to estimate *ET* at regional and even global scales. The choice of method and type of sensor depend on the amount of data required, access to the sensor images, the size of the study area, and the objectives of the study.

Allen and Tasumi [5] estimated *ET* using SEBAL and Landsat satellite images in the Bear River Basin, USA. They prepared monthly *ET* maps and provided the *ET* spatial distribution. Lysimetric ground measurements were used to validate the SEBAL model data. Bastiaanssen et al. [2] estimated ET using remote sensing and the SEBAL algorithm to study water storage plans in the Yakima basin in Washington. The results show that the annual *ET* accuracy of a large basin is ~95%. Additionally, an 85% accuracy was reported on a field scale and a ~95% accuracy was reported for seasonal estimations.

Kimura et al. [6] estimated the *ET* in the Loess Plateau of China using satellite images and the SEBAL method. They provided comparisons of models and direct measurements. James et al. [7] and Jiang et al. [8] used satellite information to detect crop uniformity, vegetation percentages, and water stress and to manage irrigation systems in India, Pakistan, Sri Lanka, Argentina, and Iran. The results showed that for 85% of the cases, the parameters estimated from remote sensing corresponded to field measurements. Kosa [9] studied the effect of temperature on the actual *ET* based on Landsat satellite imagery. The results show that the relationship between the temperature and actual evapotranspiration is in the format of the polynomial equation and that this relationship can be used to estimate actual evapotranspiration when the temperature is not known.

Merlin et al. [10] pointed to the importance of *ET* for estimating water in the soil, flood forecasting, and rainfall and predicting changes in heat waves and drought. Senay et al. [11] estimated *ET* using Landsat 8 satellite images through remote sensing in the Colorado River Basin. The results showed that there were 12% and 1.3% differences between estimated and measured values for 20-day and monthly periods, respectively. Zamani Losgedaragh and Rahimzadegan [12] evaluated SEBAL, Surface Energy Balance System (SEBS), and METRIC models for estimating evaporation from a freshwater lake in Iran. The results show the SEBAL inefficiency and the proper performance of SEBS and METRIC models for estimating evaporation from the selected water body in comparison with evaporation pan data.

Elkatoury et al. [13] evaluated and compared SEBS models for estimating regional *ET* in Saudi Arabia. They showed that the monthly *ET* results measured and calculated by SEBS models were highly correlated and consistent. Faridatul et al. [14] improved remote sensing-based ET modeling in a heterogeneous urban environment. An improved surface energy balance algorithm for urban areas (uSEBAL) was proposed to make it suitable for estimating *ET* in urban environments. Finally, the results were compared with the SEBAL algorithm. Jaafar and Ahmad [15] derived a novel time-series of field-scale actual *ET* for the Bekaa Valley in Lebanon using two one-source energy balance models, utilizing local weather data and all available original Level 1 Landsat thermal imagery and Level 2 surface reflectance products. The annual analysis showed no discernable trend in *ET* across the valley, but there was an increase in irrigated agriculture in the Orontes Basin in the last five years.

Regarding the previous studies, *ET* estimates need to be evaluated and verified for each agricultural product and environmental condition. Moreover, little research has been carried out in the field on satellite estimates of *ET* for wheat crops, which is one of the most important agricultural products of Iran. Therefore, this study aims to evaluate the efficiency of the SEBAL method using Landsat 8 satellite images (with an average spatial resolution) and vegetation indices. The main objectives of this study are (1) to evaluate spatial images of the SEBAL performance in actual *ET* validation estimation, (2) to evaluate the water requirements for the Ein Khosh Plain, and (3) to discuss the water required for irrigating the Ein Khosh Plain.

2. Materials and Methods

2.1. Study Area

As previously mentioned, the area focused on in this study is that of the Ein Khosh Plain. This is part of the Plain lands of Dehloran city in the Ilam province in western Iran. Its area is approximately 34,500 hectares and is 90 km from Dehloran city. The latitude of the region extends from 47°33′ to 47°49′ east longitude and 32°11′ to 30°25′ north latitude. This area has the longest common border between Iran and Iraq. The average height of the area is about 137 m above sea level and the average annual temperature is 25.6 °C. The average annual rainfall is 271.5 mm, and the wet season peaks in December and January. The dry period of the region is from approximately 1 April to 1 November. A map showing the region and its location in Iran and the Middle East is provided in Figure 1.

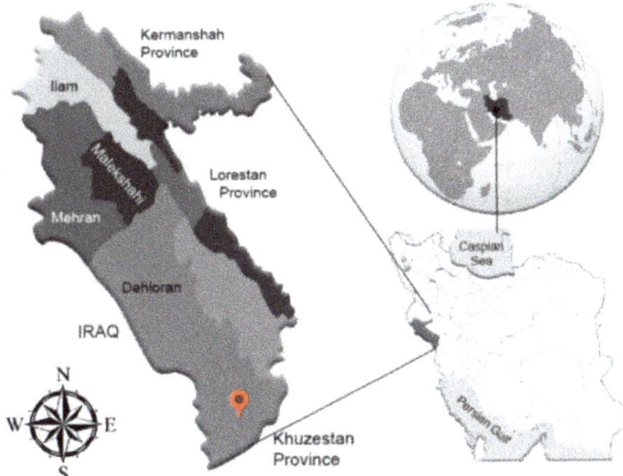

Figure 1. Map showing the location of the study area—Khosh Plain in the Ilam province of Iran.

2.2. Satellite Images

Remote sensing data obtained by satellites have the advantage of being able to provide simultaneous information over a large area. In this study, Landsat 8 satellite images processed with 11 bands were used to estimate the actual *ET* rate [16]. The 11 bands of the Landsat 8 satellite are listed in Table 1, along with wavelength and spatial resolution.

2.3. Calculation of Solar Radiation and ET

The Surface Energy Balance Algorithm for Land (SEBAL) was used to estimate the *ET* [17]. A conceptual schematic of SEBAL is presented in Figure 2.

Remote sensing was used to determine surface temperatures, estimates of net radiation (R_n), soil heat (G), latent heat (λET) fluxes, and sensible heat (H) in units of W/m^2. The latent heat flux (λET) represents the rate of heat loss from the surface due to *ET*, which was calculated for each pixel according to Equation (1):

$$\lambda ET = R_n - G - H \tag{1}$$

Table 1. Landsat 8 satellite characteristics [16].

	Bands	Wavelength (μm)	Spatial Resolution (m)
Landsat 8 Operational Land Imager (OLI) and Thermal Infrared Sensor (TIRS)	Band 1 – Coastal aerosol	0.43 – 0.45	30
	Band 2 - Blue	0.45 – 0.51	30
	Band 3 - Green	0.53 – 0.59	30
	Band 4 - Red	0.64 – 0.67	30
	Band 5 – Near Infrared (NIR)	0.85 – 0.88	30
	Band 6 –SWIR 1	1.57 – 1.65	30
	Band 7 – SWIR 2	2.11 – 2.29	30
	Band 8 - Panchromatic	0.50 – 0.68	15
	Band 9 - Cirrus	1.36 – 1.38	30
	Band 10 - Thermal Infrared (TIRS) 1	10.60 – 11.19	100
	Band 11 – Thermal Infrared (TIRS) 2	11.50 – 12.51	100

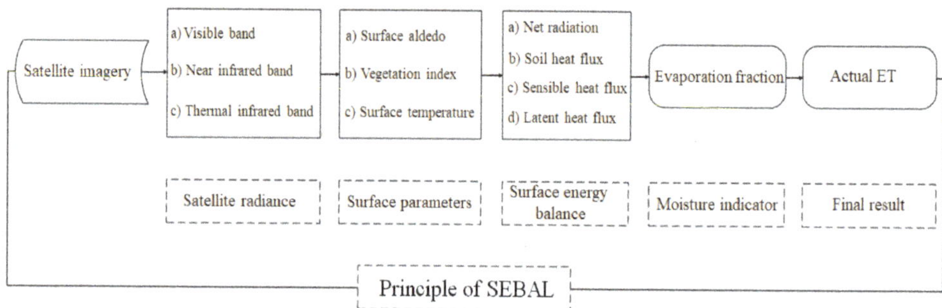

Figure 2. Principal components of the Surface Energy Balance Algorithm for Land (SEBAL) model [17].

Net radiation (R_n) is the difference between the incoming and outgoing radiative fluxes and was calculated as shown in Equation (2).

$$R_n = (1-\alpha)R_{s\downarrow} + R_{L\downarrow} - R_{L\uparrow} - (1-\varepsilon_0)R_{L\downarrow} \quad (2)$$

where $R_{s\downarrow}$ is the incoming short wavelength radiation flux, $R_{L\downarrow}$ is the incoming long wavelength radiative flux, $R_{L\uparrow}$ represents the outgoing long wave length radiative flux, α represents the surface albedo, and ε_0 is the surface emissivity. These radiant fluxes were calculated as shown in Equations (3)–(5):

$$R_{s\downarrow} = G_{sc}.\cos\theta.r.\tau_{sw} \quad (3)$$

$$R_{L\uparrow} = \varepsilon_o.\sigma.T_s^4, \quad (4)$$

$$R_{L\downarrow} = \varepsilon_\alpha.\sigma.T_\alpha^4. \quad (5)$$

Here, G_{sc} is the solar constant (1367 W/m^2), cosθ is the cosine of the solar incidence angle, r is the Earth–Sun distance, and τ_{sw} is the atmospheric transmissivity. Values for $R_{s\downarrow}$ can range from 200 to 1000 W·m^{-2}, depending on the time and location of the image and on local weather conditions. The symbol σ is the Stefan–Boltzmann constant (5.67×10^{-8} W·m^{-2}·K^{-4}), T_s is the surface temperature (K), ε_a is the atmospheric emissivity, and T_a is the atmospheric temperature (K). The following empirical equation for ε_a was applied using data from alfalfa fields in Idaho [18]:

$$\varepsilon_a = 0.85 \times (-Ln\,\tau_{sw})^{0.09}. \quad (6)$$

Here, τ_{sw} is the atmospheric transmissivity calculated assuming clear sky and relatively dry conditions. It was calculated using the elevation-based relationship of Allen et al. [19]. The soil heating

(G) is the rate of heat storage in the soil and vegetation due to conduction. The ratio of G/R_n was computed using the following empirical equation [20]:

$$G/R_n = \frac{T_S}{\alpha}(0.0038\alpha + 0.007\alpha^2)(1 - 0.98NDVI^4), \tag{7}$$

where T_s is the surface temperature (°C), α is the surface albedo, and NDVI is the normalized difference in vegetation indices between −1 and +1. Values between 0 and ~0.2 correspond to bare soil or very sparse vegetation, and NDVI > 0.2 for vegetated regions. If the NDVI value is less than zero, the surface is assumed to be water and G/R_n = 0.5. For areas where T_s < 4 °C and α > 0.45, it is assumed to be snow-covered and G/R_n = 0.5 (Allen et al. [21]). NDVI was calculated from Equation (8):

$$NDVI = \frac{R' - R}{R' + R'}, \tag{8}$$

where R is the reflectance in the red band and R' is the reflectance in the near infrared band [20]. The sensible heat flux (H) is the rate of heat loss to the air by convection and conduction (Morse et al. [22]). It was obtained from Equation (9):

$$H = \frac{\rho.C_P.(T_s - T_r)}{r_a}. \tag{9}$$

Here, ρ is the air density (kg/m^3), C_p is the specific heat of the air at a constant pressure (1004 J·kg^{-1}·K^{-1}), T_s is the surface temperature (K), T_r is the air temperature at a reference level (K), and r_a is the aerodynamic resistance to heat transport (s/m) (Allen et al. [19]). The term r_a was computed using Equation (10):

$$r_a = 1/(C_H|V|), \tag{10}$$

where C_H is the convective heat transfer coefficient and V is the wind speed at the reference level (Tasumi et al. [23]). The term ET_{inst} (instantaneous value of ET) (J/kg) is the ratio of λET to λ (the latent heat of vaporization) (J/kg) (Equation (11)):

$$ET_{inst} = 3600\frac{\lambda ET}{\lambda}. \tag{11}$$

Here, 3600 converts seconds to hours and λ is obtained according to Equation (12):

$$\lambda = 2.501 - (T_a - 273) \times 0.002361, \tag{12}$$

where T_a is the atmospheric temperature (K). The ET_{24} (actual daily ET estimation) (mm/day) is more applicable than ET_{inst}. SEBAL calculates ET_{24} assuming that the ET_rF is a 24-hour average (fixed over 24 h), according to

$$ET_{24} = ET_rF \times ET_{r-24}. \tag{13}$$

Here, ET_{r-24} is the 24-h ET_r for the day on which the image was captured; it is calculated as the sum of the hourly ET_r values for that day (Allen et al. [19]). A reference value of ET (ET_0) could be obtained by using the FAO-Penman–Monteith method (Equation (14)):

$$ET_0 = \frac{\Delta(R_n - G) + \rho C_P(e_a - e_d)/r_a}{\Delta + \gamma(1 + r_c/r_a)}, \tag{14}$$

where Δ represents the slope of the saturation vapor pressure curve (1/kPa), ρ is the atmospheric density (kg/m^3), C_P is the specific heat of the air (kJ/g°C), e_a–e_d represents the water vapor pressure deficiency (kPa), and the terms r_c and r_a are the (bulk) surface and aerodynamic resistances (s/m) and γ

psychometric constants (0.665×10^{-3} Pa) (Allen et al. [24]). The actual annual *ET* (Equation (16)) was calculated using daily *ET* data (Equation (15)) as follows:

$$ET_{period_i} = \frac{ET_{a_i}}{ET_{o_i}} \sum_{J=k}^{l} ET_{oj}, \quad (15)$$

$$ET_{annual} = \sum ET_{period_i}. \quad (16)$$

Here, ET_{ai} is the actual *ET* obtained from the images on the same day of the image being taken (*i*th day of the year) (mm), ET_{oi} is the reference *ET* from the FAO-Penman–Monteith equation (also for the *i*th day of the year) (mm), ET_{oj}, is the ET related to the number of days in the period of image *i* that varies from the *k*th to the *l*th day of the year, and *j* represents the number of days. The last term, ET_{annual}, is the actual annual ET obtained from the sum of the $ET_{period}i$ (mm).

To calculate the annual *ET*, Landsat 8 satellite images were used during cropping and harvesting times and in clear sky conditions. Wheat is planted in autumn (late November) in the Ein Khosh Plain. The crop harvest occurs at the end of June. Images for this time range were obtained and ENVI-4.2 software was used to process and prepare those images for the SEBAL algorithm. In addition, REF-ET was used to calculate the reference *ET*. The dates of the images are presented in Table 2.

Table 2. Dates of the images used in the present study.

Satellite	Date of Pictures (AD)
Landsat 8	11-12-2014
Landsat 8	10-1-2015
Landsat 8	29-2-2015
Landsat 8	27-3-2015
Landsat 8	20-4-2015
Landsat 8	17-5-2015
Landsat 8	04-6-2015

The statistical criteria of the Mean Bias Error (MBE), Root Mean Square Error (RMSE), Mean Absolute Percentage Error (MAPE), and correlation coefficient (R^2) were used to evaluate the model. These metrics were calculated as follows:

$$MBE = \frac{1}{N} \sum_{i=1}^{N} (O_i - P_i) \quad (17)$$

$$RMSE = \sqrt{\frac{\sum_{i=1}^{N} (O_i - P_i)^2}{N}} \quad (18)$$

$$MAPE = \left[\frac{1}{N} \sum_{i=1}^{N} \left| \frac{O_i - P_i}{O_i} \right| \right] \times 100 \quad (19)$$

Here, O_i represents the observed values of the FAO-Penman–Monteith equation as the standard model; P_i represents the estimated values from the SEBAL algorithm; and $\overline{O_i}$ and $\overline{P_i}$ are the mean values from the FAO Penman–Monteith model and SEBAL, respectively.

3. Results and Discussions

3.1. The Net Radiation (R_n), Soil Heat (G), and Sensible Heat (H) Fluxes

Figure 3 shows the net radiation (R_n), soil heat (*G*), and sensible heat (*H*) fluxes for the Ein Khosh Plain. According to the map and the R_n values, the maximum radiation occurs in areas of vegetation growth and minimum values occur in areas without vegetation. Additionally, it was observed that the values of *G* in vegetated areas were in the range of 0.05 to 0.15, which is the rate of conduction heat transfer within the soil. The sensible heat results are also plotted in Figure 3.

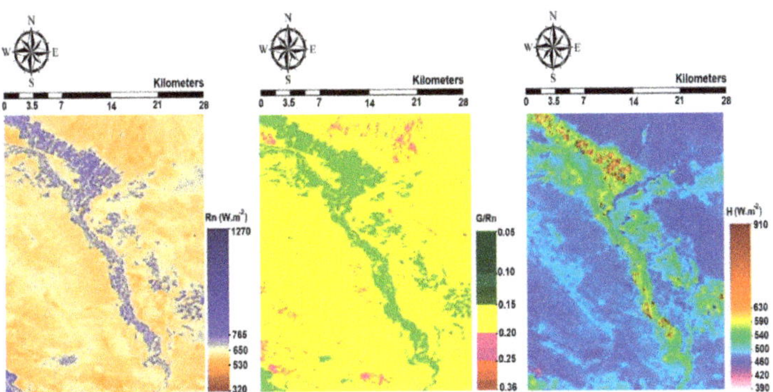

Figure 3. Net radiation (Rn), soil heat (G), and sensible heat (H) fluxes for the Ein Khosh Plain.

3.2. Evaluation of SEBAL's Performance in Actual ET Validation Estimation

After estimating R_n, *G*, and *H*, it was possible to determine the daily *ET* rates and the results could be compared with calculations using the FAO-Penman–Monteith equation as the reference method. The nearest station with daily and hourly data is located at 32° 15′ north and 48° 24′ east. Using 3-h station data, the daily *ET* was calculated and the water requirements were obtained during the growth period. The *ET* rates obtained from the FAO-Penman–Monteith and SEBAL methods are presented in Table 3. Furthermore, Figure 4 shows a comparison of the actual *ET* values calculated by SEBAL with the FAO-Penman–Monteith model values.

Table 3. Evaluation of evapotranspiration (*ET*) of FAO-Penman–Monteith and SEBAL methods.

Date of Pictures (AD)	FAO-Penman-Monteith	SEBAL
	ET_0 (mm/day)	ET_0 (mm/day)
11-12-2014	3.87	3.51
10-01-2015	4.21	4.53
29-02-2015	4.89	5.01
27-03-2015	5.73	5.16
20-04-2015	8.22	8.44
17-05-2015	9.54	8.98
04-06-2015	9.51	8.74

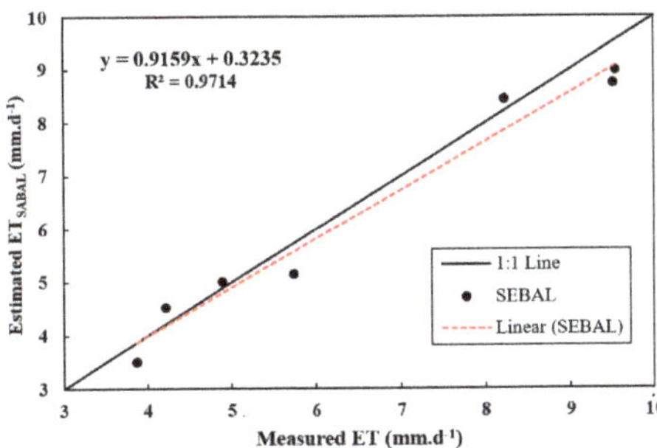

Figure 4. Comparison of the actual ET values calculated using SEBAL with the FAO-Penman–Monteith method.

As is shown in Table 3 and Figure 4, the values of *RMSE, MAPE*, and *MBE* were 0.466, 2.9%, and 0.222 mm/day, respectively, with a correlation coefficient of 0.97, indicating that SEBAL's accuracy is sufficient for estimating the actual *ET*. It can be seen that remote sensing is an efficient and effective way to estimate *ET* on large scales, especially in areas where meteorological data is not available. The actual daily *ET* rates in the Ein Khosh Plain for different months are shown in Figure 5.

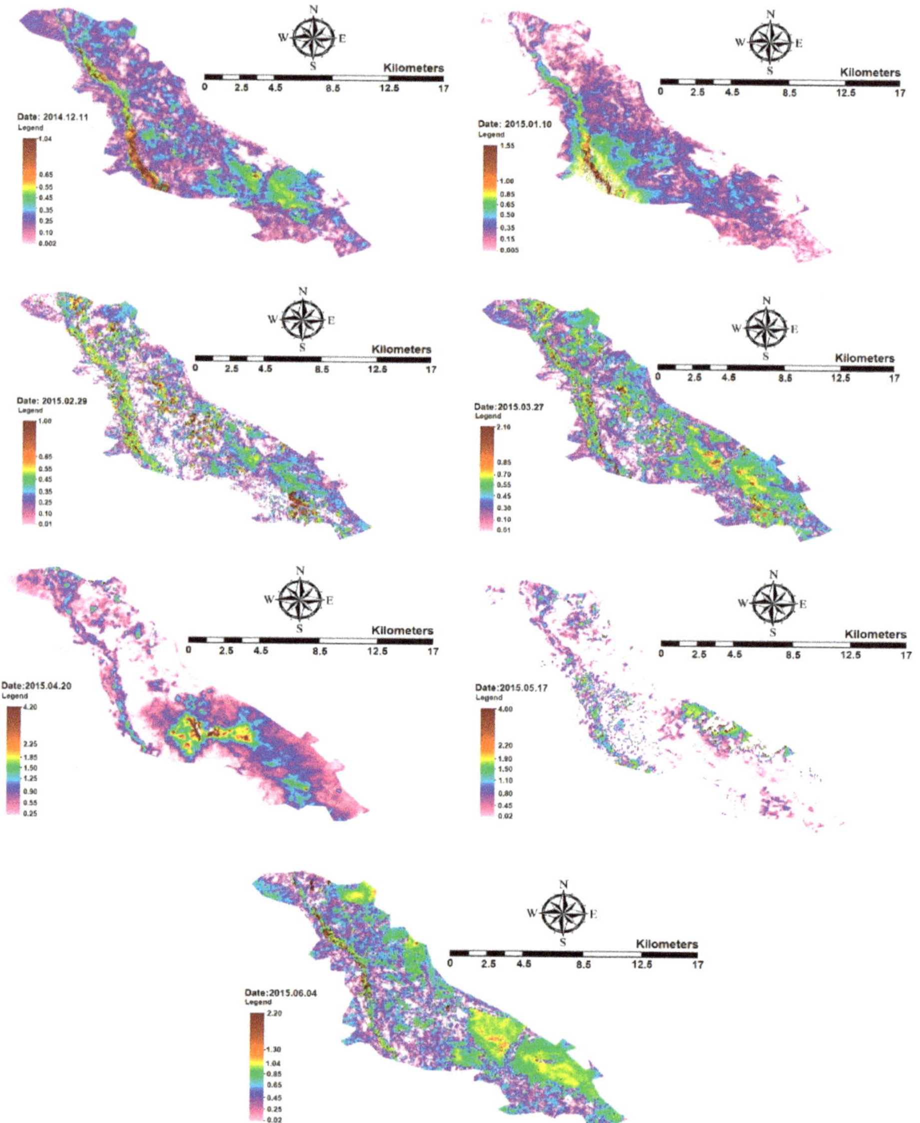

Figure 5. Actual daily ET rate in the Ein Khosh Plain during different months.

Figure 6 shows the actual annual *ET* in the Ein Khosh Plain. It can be observed that the maximum annual *ET* occurs in areas with a high density of vegetation.

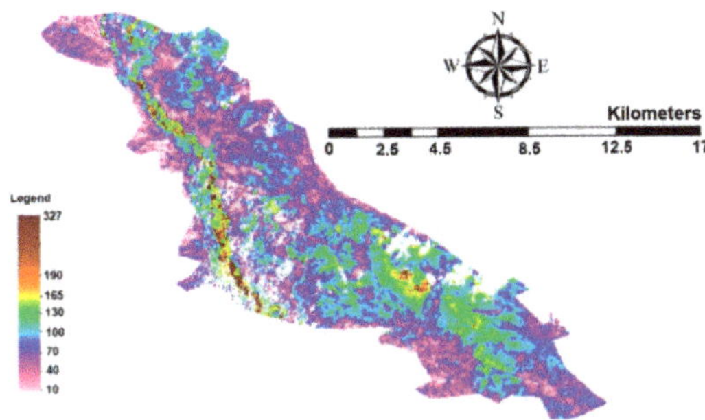

Figure 6. Actual annual ET in the Ein Khosh Plain (mm/year).

3.3. Estimation of the Water Requirement for the Ein Khosh Plain

In order to estimate the water requirement of the Ein Khosh Plain, the study areas were classified into three categories: cultivated, not cultivated/fallow, and rangeland/wasteland. Images were classified using an object-oriented method and eCognation software was employed. Each image was classified within the software based on a threshold value. Image segmentation was performed in order to subdivide the overall image into multiple non-overlapping parts (Zoleikani et al. [25]). An example outcome of this process is provided in Figure 7.

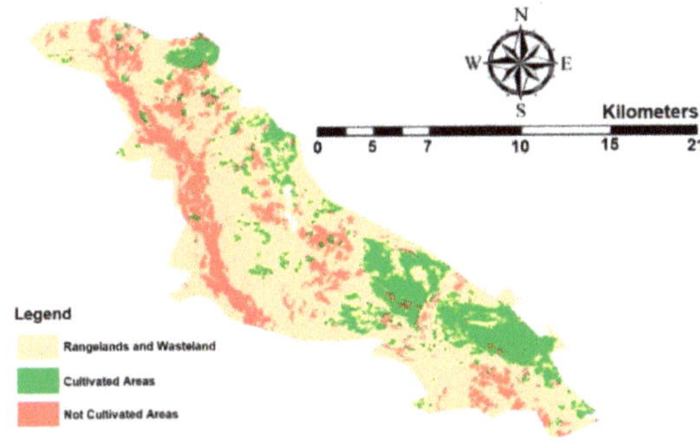

Figure 7. Classified image of the Ein Khosh Plain (28 May).

The area of each category of the Ein Khosh Plain is presented in Table 4.

Table 4. Area of use for the Ein Khosh Plain.

Plain	Uses	Area (km^2)	Average Actual Annual ET
	Rangeland and wasteland	239.99	75
Ein Khosh	Cultivated	62.89	121
	Not cultivated	60.23	113

The evaluation of *ET* in the Ein Khosh Plain shows that the 121 mm average *ET* is related to agricultural lands. The total area of the plain is 363.11 km^2. Taking into account the area of the Ein Khosh Plain and the cultivated land, the cultivation density is 17.21%. The normalized difference vegetation index (NDVI) was calculated according to Equation (7) and is shown in Figure 8.

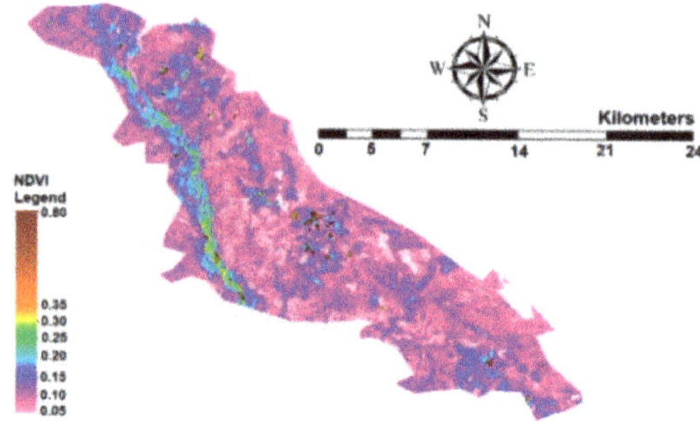

Figure 8. A normalized difference vegetation index (NDVI) map of the Ein Khosh Plain (28 May).

Figure 8 highlights that the NDVI values between 0.2 and 0.5 correspond to agricultural cover, representing a small area of the Ein Khosh Plain. To study water management in these periods, first, the amount of water consumed per hectare for each agricultural land was calculated. Next, the amount of water needed for irrigation was estimated based on the amount of rainfall. Table 5 shows the amount of water consumed per hectare for the Ein Khosh Plain agricultural lands.

Table 5. Calculation of the amount of water consumed per hectare of Ein Khosh Plain lands.

Image Date	Study Period	Uses	Area (hr)	Amount of Water Required (mm/hr)
11-12-2014	1st —18 November to 25 December	agricultural lands	6318.5	127.11
10-01-2015	2nd—26 December to 15 February	agricultural lands	6318.5	177.62
29-02-2015	3rd—16 February to 15 March	agricultural lands	6318.5	19.47
27-03-2015	4th—16 March to 13 April	agricultural lands	6318.5	231.23
20-04-2015	5th—14 April to 9 May	agricultural lands	6318.5	227.22
17-05-2015	6th—10–25 May	agricultural lands	6318.5	143.57
04-06-2015	7th—26 May to 10 June	agricultural lands	6318.5	24.53

According to Table 5, it could be observed that the highest amount of water required was found in the fourth period (March 16 to April 13), with a value of 231.23 mm/hr, and the lowest was found in the third period (February 16 to March 15), with a value of 19.47 mm/hr, for agricultural land use.

3.4. Water Required for Irrigation of Ein Khosh Plain in Each Period

Figure 9 shows the monthly irrigation requirement of the Ein Khosh Plain for different months of the year.

It could be observed that the rainfall in December is higher than the water requirements and as a consequence, there is an excess of rainwater. The dark green areas correspond to surplus water in agricultural land and the value is ~15 mm. Rainfall in January exceeds *ET* and there is no need to irrigate the agricultural areas. Despite the low *ET* in February, rainfall is extremely low during this month and little irrigation water is required. The yellow and green regions correspond to areas with 0–5 and 5–20 mm water deficiencies. Additionally, March's rainfall corresponds to *ET* losses and consequently, does not require irrigation. By comparing the April irrigation requirement map with the NDVI map, the denser parts of the vegetation appear to require about 50 mm of irrigation water. Moreover, according to Figure 9 and the map of NDVI, the maximum water requirement in May is 70 mm for dense vegetation. The yellow areas on the map correspond to water deficiencies of up to 15 mm. Due to the lack of rainfall in June, the rainfall in the whole area is less than the amount required. The maximum water deficit during this month is 70 mm, which must be supplied using other water sources. Figure 10 shows the annual irrigation requirements for the Ein Khosh Plain. A careful inspection of Figure 10 and comparison with Figure 6 reveals the irrigation requirement for the dark green parts of the map, which are deficient by up to 20 mm during the crop season. These spots are very small and not noticeable. This indicates that the land does not require irrigation. Irrigation may occur during the growing season due to a lack of rainfall, but this requirement is not seen throughout the growing season. The rainfall in the region seems to be responsive to the amount of water required in the Ein Khosh Plain.

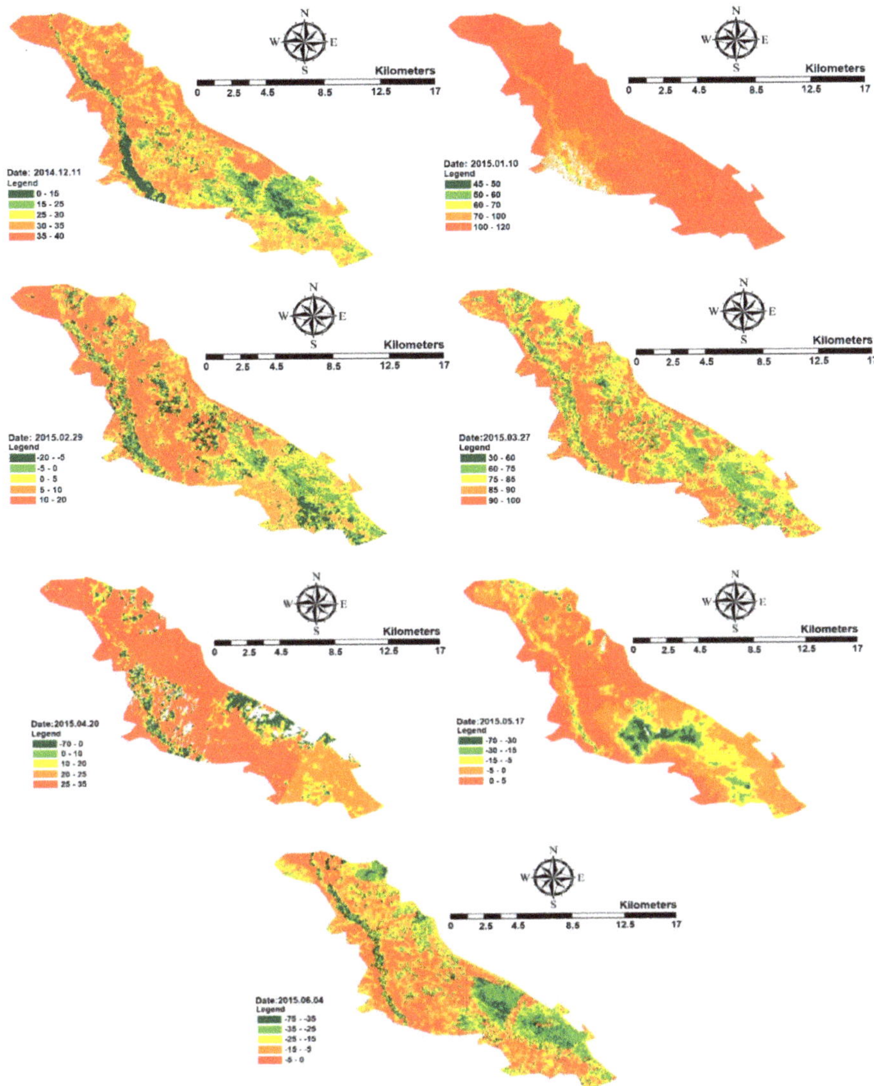

Figure 9. Monthly irrigation requirement of the Ein Khosh Plain for different months of the year.

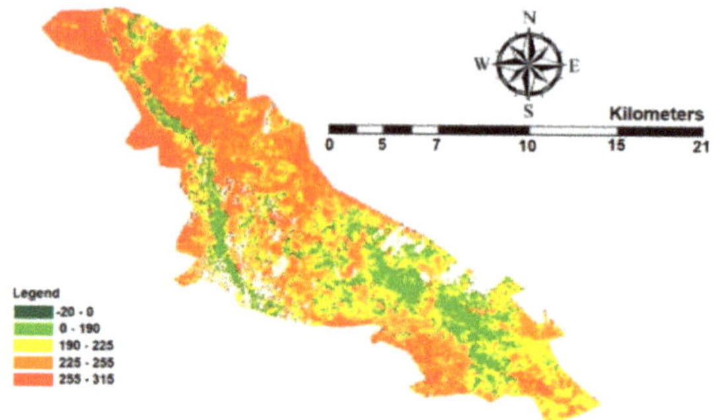

Figure 10. Annual irrigation requirement of the Ein Khosh plain.

4. Conclusions

Proper estimations of the plant *ET* and water requirements of plants are very important for improving water management and increasing the water consumption efficiency. In this regard, satellite *ET* estimation models such as SEBAL can be useful. Of course, the efficiency of this model is different in various climates and crops. Therefore, the purpose of this study was to evaluate the efficiency of the SEBAL algorithm for wheat crops, which is one of the most important agricultural products of Iran. Therefore, the *ET* rate was estimated by remote sensing and analyzed using the SEBAL algorithm for the Ein Khosh Plain in the Ilam province in Iran. The net radiation (R_n), soil heat (G), and sensible heat (H) fluxes and NDVI were plotted and analyzed. The evaluation of SEBAL with the FAO-Penman–Monteith method as a reference showed that the values of RMSE, MAPE, and MBE were 0.466, 2.9%, and 0.222 mm/day, respectively, with a correlation coefficient of 0.97. It was proven that SEBAL has a sufficient accuracy for estimating the actual *ET*. The results of the SEBAL algorithm are as follows:

- The rainfall rate in the Ein Khosh Plain, except for the last month of cultivation with very low rainfall, meets water-use requirements, except for late May and early June. Despite the lower *ET* rate for wheat in the last month, there is a need for irrigation during this month;
- An evaluation of irrigation requirements using monthly rainfall data showed that the Ein Khosh Plain in March (the rainfall corresponds to the *ET* rate for wheat corps), which displays the maximum *ET*, has no deficiency of rainfall. Some parts of the plain in several months, such as April and May, expect a rainfall value of up to 50 and 70 mm, respectively;
- While the total area of the plain is equal to 363.11 km^2, only 17.21% of the region is cultivated. Given that the average *ET* rate is 121 mm in the agricultural lands, a maximum of 20 mm of irrigation is required;
- During the wheat plant growth periods, the highest amount of water required was found in the fourth period (March 16 to April 13), with a value of 231.23 mm/hr, and the lowest was found in the third period (February 16 to March 15), with a value of 19.47 mm/hr, for agricultural land use.

In this study, the SEBAL model estimated the actual *ET* of the wheat crops with a sufficient accuracy compared with the FAO-Penman–Monteith method by using the minimum meteorological data. Overall, the results proved that the SEBAL algorithm can be an appropriate method for estimating the wheat *ET* and can be used as an efficient tool for managing water resources in wheat farms, forestry projects, etc.

Not only does the use of this technique help preserve water resources, but it also reduces water use during times when water is not needed. This guidance is important because excess water will increase the soil-water content and consequently the pore pressure. Excessive water can result in soils becoming unstable, producing landslides or other unintended consequences.

Author Contributions: Conceptualization, A.G., M.D., M.S., J.A., and B.C.; methodology, A.G., M.D., and M.S.; software, A.G, M.D, M.S., and B.C.; validation, A.G., M.D., and B.C.; formal analysis, A.G. and M.D.; investigation, A.G., M.D., M.S., and B.C.; writing—original draft preparation, A.G., M.D., M.S., and J.A.; writing—review and editing, A.G., M.D., and J.A.; supervision, J.A. and B.C.; project administration, A.G., J.A., and B.C. All authors have read and agreed to the published version of the manuscript.

Funding: This research received no external funding.

Conflicts of Interest: The authors declare no conflicts of interest.

References

1. Li, Y.L.; Cui, J.Y.; Zhang, T.H.; Zhao, H.L. Measurement of evapotranspiration of irrigated spring wheat and maize in a semi-arid region of north China. *Agric. Water Manag.* **2003**, *61*, 1–12. [CrossRef]
2. Bastiaanssen, W.G.M.; Noordman, E.J.M.; Pelgrum, H.; Davids, G.; Thoreson, B.P.; Allen, R.G. SEBAL model with remotely sensed data to improve water-resources management under actual field conditions. *J. Irrig. Drain. Eng.* **2005**, *131*, 85–93. [CrossRef]
3. Zhang, X.C.; Wu, J.W.; Wu, H.Y.; Li, Y. Simplified SEBAL method for estimating vast areal evapotranspiration with MODIS data. *Water Sci. Eng.* **2011**, *4*, 24–35.
4. Trezza, R.; Allen, R.G.; Tasumi, M. Estimation of actual evapotranspiration along the Middle Rio Grande of New Mexico using MODIS and Landsat imagery with the METRIC model. *Remote Sens.* **2013**, *5*, 5397–5423. [CrossRef]
5. Allen, R.G.; Tasumi, M. Appendix B: Algorithm for applying SEBAL to sloping mountainous areas. In *Application of the SEBAL Methodology for Estimating Consumptive Use of Water and Stream Flow Depletion in the Bear River Basin of Idaho Through Remote Sensing*; Idaho Department of Water Resource: Boise, ID, USA, 2000.
6. Kimura, R.; Bai, L.; Fan, J.; Takayama, N.; Hinokidani, O. Evapo-transpiration estimation over the river basin of the Loess Plateau of China based on remote sensing. *J. Arid Environ.* **2007**, *68*, 53–65. [CrossRef]
7. James, K.A.; Stensrud, D.J.; Yussouf, N. Value of real-time vegetation fraction to forecasts of severe convection in high-resolution models. *Weather Forecast.* **2009**, *24*, 187–210. [CrossRef]
8. Jiang, L.; Kogan, F.N.; Guo, W.; Tarpley, J.D.; Mitchell, K.E.; Ek, M.B.; Ramsay, B.H. Real-time weekly global green vegetation fraction derived from advanced very high resolution radiometer-based NOAA operational global vegetation index (GVI) system. *J. Geophys. Res. Atmos.* **2010**. [CrossRef]
9. Kosa, P. The effect of temperature on actual evapotranspiration based on Landsat 5 TM satellite imagery. *Evapotranspiration* **2011**, *56*, 209–228.
10. Merlin, O.; Chirouze, J.; Olioso, A.; Jarlan, L.; Chehbouni, G.; Boulet, G. An image-based four-source surface energy balance model to estimate crop evapotranspiration from solar reflectance/thermal emission data (SEB-4S). *Agric. For. Meteorol.* **2014**, *184*, 188–203. [CrossRef]
11. Senay, G.B.; Friedrichs, M.; Singh, R.K.; Velpuri, N.M. Evaluating Landsat 8 evapotranspiration for water use mapping in the Colorado River Basin. *Remote Sens. Environ.* **2016**, *185*, 171–185. [CrossRef]
12. Losgedaragh, S.Z.; Rahimzadegan, M. Evaluation of SEBS, SEBAL, and METRIC models in estimation of the evaporation from the freshwater lakes (Case study: Amirkabir dam, Iran). *J. Hydrol.* **2018**, *561*, 523–531. [CrossRef]
13. Elkatoury, A.; Alazba, A.A.; Mossad, A. Estimating evapotranspiration using coupled remote sensing and three SEB models in an arid region. *Environ. Process.* **2019**, *7*, 109–133. [CrossRef]
14. Faridatul, M.I.; Wu, B.; Zhu, X.; Wang, S. Improving remote sensing based evapotranspiration modelling in a heterogeneous urban environment. *J. Hydrol.* **2020**, *581*, 124405. [CrossRef]
15. Jaafar, H.H.; Ahmad, F.A. Time series trends of Landsat-based ET using automated calibration in METRIC and SEBAL: The Bekaa Valley, Lebanon. *Remote Sens. Environ.* **2020**, *238*, 111034. [CrossRef]
16. Oguz, H. Calculating actual evapotranspiration in Adana, Turkey using Landsat 8 imagery with the METRIC model. In Proceedings of the 1st International Symposium of Forest Engineering and Technologies, Bursa, Turkey, 2–4 June 2016.
17. Bastiaanssen, W.G.M.; Menenti, M.; Feddes, R.A.; Holtslag, A.A.M. The surface energy balance algorithm for land (SEBAL): Part 1 formulation. *J. Hydrol.* **1998**, *212*, 198–212. [CrossRef]

18. Bastiaanssen, W.G. *Regionalization of Surface Flux Densities and Moisture Indicators in Composite Terrain a Remote Sensing Approach Under Clear Skies in Mediterranean Climates*; SC-DLO: Wageningen, The Netherlands, 1995; p. 271.
19. Allen, R.; Tasumi, M.; Trezza, R.; Waters, R.; Bastiaanssen, W. SEBAL (Surface Energy Balance Algorithms for Land). In *Advance Training and User's Manual*, version 1; Idaho Implementation: Sagle, ID, USA, 2002; Volume 1, p. 97.
20. Bastiaanssen, W.G. SEBAL-based sensible and latent heat fluxes in the irrigated Gediz Basin, Turkey. *J. Hydrol.* **2000**, *229*, 87–100. [CrossRef]
21. Allen, R.; Irmak, A.; Trezza, R.; Hendrickx, J.M.; Bastiaanssen, W.; Kjaersgaard, J. Satellite-based ET estimation in agriculture using SEBAL and METRIC. *Hydrol. Process.* **2011**, *25*, 4011–4027. [CrossRef]
22. Morse, A.; Allen, R.G.; Tasumi, M.; Kramber, W.J.; Trezza, R.; Wright, J.L. *Final Report: Application of the SEBAL Methodology for Estimating Evapotranspiration and Consumptive Use of Water Through Remote Sensing*; Idaho Department of Water Resources, University of Idaho, Department of Biological and Agricultural: Moscow, ID, USA, 2000; Volume 107.
23. Tasumi, M.; Trezza, R.; Allen, R.G.; Wright, J.L. US Validation Tests on the SEBAL Model for Evapotranspiration via Satellite. In Proceedings of the 54th IEC Meeting of the International Commission on Irrigation and Drainage (ICID) Workshop Remote Sensing of ET for Large Regions, Montpellier, France, 14–17 September 2003.
24. Allen, R.B.; Pereira, L.S.; Raes, D.; Smith, M.S. Crop evapotranspiration (guidelines for computing crop water requirements). *FAO Irrig. Drain. Pap.* **1998**, *56*, 300.
25. Zoleikani, R.; Zoej, M.V.; Mokhtarzadeh, M. Comparison of pixel and object oriented based classification of hyperspectral pansharpened images. *J. Indian Soc. Remote Sens.* **2017**, *45*, 25–33. [CrossRef]

Technical Note

Determination of the Probabilities of Landslide Events—A Case Study of Bhutan

Raju Sarkar [1,2,*] and Kelzang Dorji [1]

1 Center for Disaster Risk Reduction and Community Development Studies, Royal University of Bhutan, Rinchending 21101, Bhutan; kelzangdorji.cst@rub.edu.bt
2 Department of Civil Engineering, Delhi Technological University, Bawana Road, Delhi 110042, India
* Correspondence: rajusarkar.cst@rub.edu.bt

Received: 30 May 2019; Accepted: 13 June 2019; Published: 16 June 2019

Abstract: Landslides have been and are prominent and devastating natural disasters in Bhutan due to its orography and intense monsoonal rainfall. The damage caused by landslides is huge, causing significant loss of lives, damage to infrastructure and loss of agricultural land. Several methods have been developed to understand the relationship between rainfall and landslide incidences. The most common method to understand the relationship is by defining thresholds using empirical methods which are expressed in either intensity-duration or event rainfall-duration terms. However, such thresholds determine the results in a binary form which may not be useful for landslide cases. Apart from defining thresholds, it is significant to validate the results. The article attempts to address both these issues by adopting a probabilistic approach and validating the results. The region of interest is the Chukha region located along the Phuentsholing-Thimphu Highway, which is a significant trade route between neighbouring countries and the national capital Thimphu. In the present study, probabilities are determined by Bayes' theorem considering rainfall and landslide data from 2004 to 2014. Singular (rainfall intensity, rainfall duration and event rainfall) along with a combination (rainfall intensity and rainfall duration) of precipitation parameters were considered to determine the probabilities for landslide events. A sensitivity analysis was performed to verify the determined probabilities. The results depict that a combination of rainfall parameters is a better indicator to forecast landslides as compared to single rainfall parameter. Finally, the probabilities are validated using landslide records for 2015 using a threat score. The validation signifies that the probabilities can be used as the first line of action for an operational landslide warning system.

Keywords: probabilistic method; Bhutan Himalayas; shallow landslides

---

1. Introduction

Landslides are the most common and one of the prominent natural hazards in steep terrain affecting human lives and property, and often leading to loss of lives. In [1] the study of a global database of landslide occurrences between 2004–2016 showed that 75% of landslides occurred in Asia, with significant occurrences in the Himalayan arc. The study also showed that the majority of the landslides are shallow landslides and are triggered by rainfall. The high or low intensity reinforced with continuous occurrence of rainfall leads to an increase in pore water pressure due to the constant saturation of soil mass, which leads to landslides [2]. The incidence of landslides is likely to increase in the future due to a growing population, increased construction activities, and exploitation of natural resources. To understand the phenomena of shallow landslides, several models were developed and used in several parts of the world. These models can be broadly categorised as empirical or physical models. Physical models require the use of several parameters involving slope-stability modelling and acquiring the data required to carry out such an analysis are usually challenging [3]. Therefore, the most used technique is the empirical model which uses the observed rainfall and landslide data which are analyzed statistically without considering the physical processes governing the landslide [4–6]. The key in utilising such a model is the availability and quality of data which reflects in the results [7]. The performance of the threshold-based models can decrease significantly with a small change in landslide data [8]. References [9–11] highlighted that definition of rainfall event and information on landslide data had a significant impact on threshold estimation which often leads to a high number of false alarms for application in an early warning system. Recent developments have led to reproducible conditions for landslide occurrences in the region of interest [7,12,13]. However, one major drawback for using empirical models is the outcome of an absolute value either in terms of rainfall intensity or cumulative event rainfall [14]. Such a result may not always help in setting up an early warning system since a small variation in calculation can lead to catastrophes. To overcome the use of empirical models, probabilistic model using Bayes' theorem was developed by [14] is an efficient choice and the same was applied to the Emilia-Romagna region in Italy. Thereafter, similar work was carried out in several places (e.g., Ha Giang region, Vietnam [15], Sierra Norte De Puebla, Mexico [16], Kalimpong, India [17,18], Chibo, India [19]. The use of probabilistic techniques has also been carried out for severe precipitation conditions [20,21].

The key to understand the thresholds or any analysis is its validation, [6] reported that out of 115 rainfall thresholds determined between 2008–2016, only 69 provided validation of their work. For validation work, only 33% of the work was carried out using an independent dataset. The most common technique for validation is skill score [8,19] or by comparing two different threshold models [22].

The Bhutan Himalayas region has been adversely affected by landslides which are caused by heavy monsoon precipitation. In the case of Bhutan, the literature available on rainfall-induced landslides is very sparse, with only one work showing the determination of event rainfall-rainfall duration (ED) thresholds [13]. The country also suffers from the absence of an operational early warning system and therefore understanding the relation using different models is very critical in developing such an early warning system [13]. This paper determines and validates the probabilities of landslide incidents using the Bayes' theorem by utilizing the available landslide and rainfall data for the Chukha region. The validation of the thresholds has been determined using a threat score which analyses the percentage of correctly predicted landslide events. The probabilities were calculated using the dataset for 11 years (2004–2014) and validated using landslide data of 2015.

2. Study Area

Bhutan lies in the eastern part of Himalaya and is surrounded by India to the east, west and south and by the Tibetan plateau to the north. The elevation of the country varies from 100 m to 7500 m [23] with elevations in Chukha region ranging from 1000 m to 4200 m. Figure 1 shows the present study area—the Chukha region. A part of the Phuentsholing-Thimphu highway (also called Asian Highway), which connects Thimphu, the capital of Bhutan, with neighbouring countries, lies in the study region. Said highway is a major trade route with neighbouring countries and is frequently blocked by landslide debris, causing severe economic damage every year.

The landslides in the region are primarily caused due to heavy monsoons and are shallow in nature. The increase in landslide events is frequently activated by toe cutting of the slope and blasting activities for infrastructure development. The increase in anthropogenic activities has escalated deforestation leading to slope instability. The slope failures along the roads usually block the road leading to huge logistic issues and economic loss.

The region comprises of active sedimentary and metasedimentary rocks, gneiss, schist, quartzite, and limestone and is underlain mostly with schistose rocks. The soils in the area consist of weaker phyllites which make the soil texture very fine and the slopes unstable. Geologically, the region is surrounded by major tectonostratigraphic units and structures are Shivalik ranges, Main Boundary Thrust (MBT), the Lesser Himalayan Sequence (LHS), MCT, HHC, and STD [24]. The major parts of the area are underlain by Baxa formation group where the thrust fault, strike and dip of bedding and foliation were more concentrated.

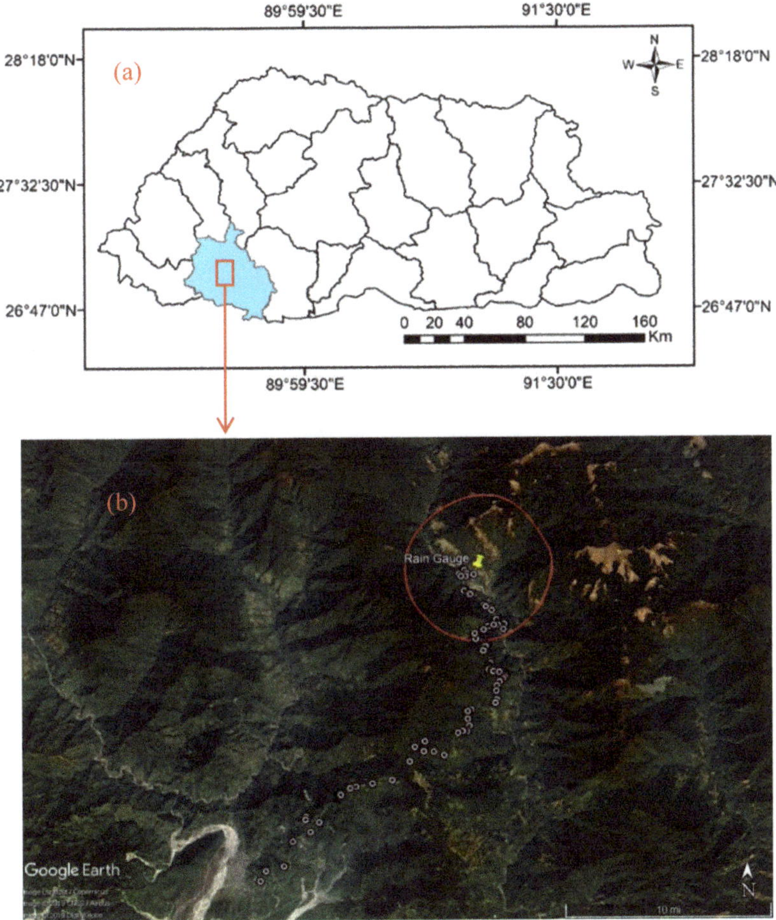

Figure 1. Location of (**a**) Bhutan (**b**) spatial distribution of landslides considered for analysis.

The slope of the region varies from 15°–75° with a majority of the area (35%) lying in the moderate steep slope (30°–45°). The presence of weak geology and weathered rocks makes the region highly susceptible to landslides. Figure 2 shows typical damages caused by landslides in the area.

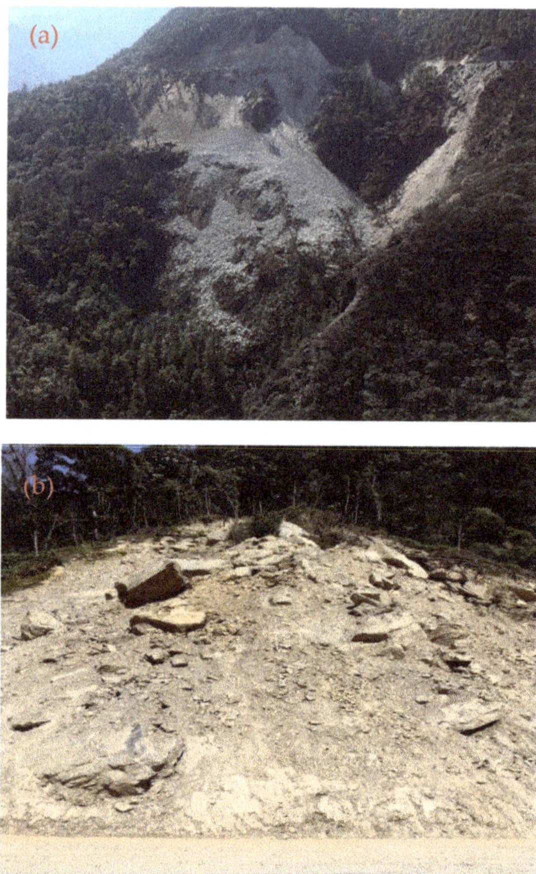

Figure 2. Landslide damages (**a**) near Chukha Dzongkhag (**b**) along the Phuentsholing–Thimphu highway.

3. Methodology

The methodology involves the use of Bayes' theorem which uses either single or multiple rainfall characteristics to determine one-dimensional and two-dimensional probabilities respectively. The technique can be applied for degrees of variables depending on the quality and quantity of rainfall and landslide data. A detailed explanation of the method along with a solved example has been presented in [14].

3.1. One-Dimensional Bayesian Probability

This type of probability approach involves the use of a single rainfall parameter to determine the landslide probability and is defined as:

$$P(X|Y) = \frac{P(Y|X) \cdot P(X)}{P(Y)} \quad (1)$$

where P(Y|X) = precipitation probability of magnitude Y for landslide occurrence, P(X) = landslide probability regardless of precipitation event of magnitude X, P(Y) = precipitation probability of magnitude X, regardless of landslide occurrence and P(X|Y) = landslide probability for precipitation event of magnitude Y.

If the number of rainfall events is M_R, number of landslide events being M_A, number of precipitation events of degree B be M_B and the number of precipitation events leading to slide initiation be $M_{(B|A)}$, then Equation (1) reduces to the computation of the following frequencies:

$$P(X) \approx M_A/M_R \quad (2)$$

$$P(Y) \approx M_B/M_R \quad (3)$$

$$P(Y|X) \approx M_{(B|A)}/M_A \quad (4)$$

The parameters being used for the analysis would depend on the key reasons for slide initiation for a region. The variables used in the present are cumulative event rainfall, rainfall duration, and rainfall intensity.

3.2. Two-Dimensional Bayesian Probability

The two-dimensional Bayesian probability determines the conditional probability of landslide event due to the common occurrence of two rainfall parameters:

$$P(X|Y, Z) = \frac{P(Y, Z|X) \cdot P(X)}{P(Y|Z)} \quad (5)$$

where Y, Z denotes the combined probability of having a certain or range of value of any two variables. If, Y equals rainfall intensity and Z is rainfall duration, the likelihood of landslide occurrence due to a rainfall event of given duration and intensity is expressed using Equation (5). The model allows any set of rainfall parameters to be used to calculate thresholds and its significance can be compared with prior probability [14]. Also, it can also be used for n-variables by modifying the equation accordingly like understanding the effect of a combination of rainfall intensity, duration, event rainfall and antecedent rainfall on landslide incidence.

4. Data

For the present study, the daily rainfall data was collected from rain gauge installed at Chuka and maintained by the National Center for Hydrology and Meteorology of the Royal Government of Bhutan. The daily rainfall data were collected for 1 January 2004 to 31 December 2014. The variation of the rainfall during the entire duration has been depicted in a box and whisker plot (Figure 3). The average rainfall during 2004–2014 was 1663.4 mm with maximum rainfall of 2926.6 mm occurring in 2012. The monsoonal rain (June–August) contributed about 80% of annual rainfall.

The landslide data for the study was provided by the Border Roads Organization, (Project DANTAK), Government of India, which maintains the Phuentsholing-Thimphu highway. The temporal distribution of landslides indicates that majority of the landslides initiate during the monsoonal period, and the spatial distribution depicts that landslide events occur along the highway. The spatial distribution of landslides is depicted in Figure 1b and involved materials are quartzite, phyllite, gneiss. Landslide event has been identified as only those landslides mapped as single points which were triggered due to rainfall. Also, if multiple events occurred on a day it was counted as a single event. The effect of the spatial distribution of landslides was considered by drawing a buffer radius of 5 km around the rain gauge [7]. A large search radius is usually required when the study area suffers from low rain gauge density [13]. A total of 123 landslide events occurred during 2004–2014. Considering the above mentioned factors, the number of landslide events was reduced to 66.

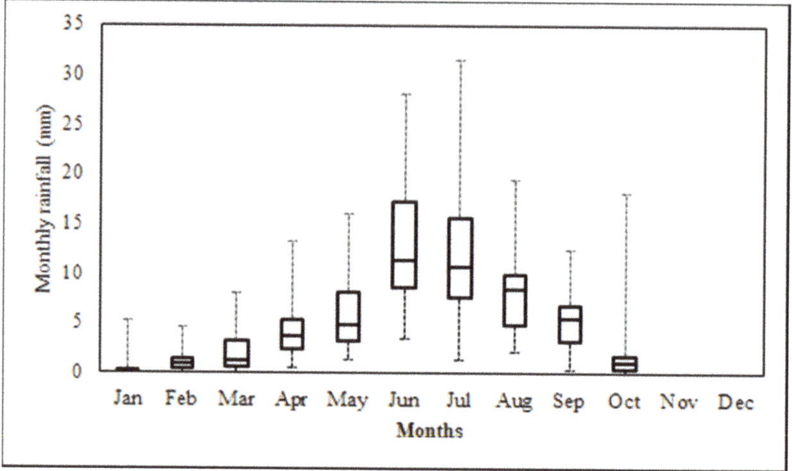

Figure 3. Box and whisker plots showing the annual variation of monthly rainfall measures in the study area for the 11-year period 2004–2014.

5. Results

The key to determine any thresholds is to define rainfall and landslide event. In this study, rainfall event is defined as the total number of consecutive days of rainfall. The rainfall intensity (mm/day) is determined by dividing the corresponding total rainfall (mm) by the rainfall duration (days). The landslide event was described as the single rainfall triggered slide activity. The landslide event for analysis includes the spatiotemporal distribution, i.e., dates and coordinates of each landslide. The total number of rainfall and landslide events determined was 480 and 66, respectively. The results of one-dimensional Bayesian probability are depicted in Figure 4a–f.

The results show that probability reaches the highest value of 0.5 for cumulative event rainfall of 300–350 mm and a probability of 0.33 for rainfall duration of 9 days. In case of rainfall intensity, the likelihood is 0.5 for the intensity of 80–90 mm/day and 0.227 for the intensity of 30–40 mm/day. This shows that heavy rainfall as well as slow and continuous rainfall can lead to landslide incidence in the region.

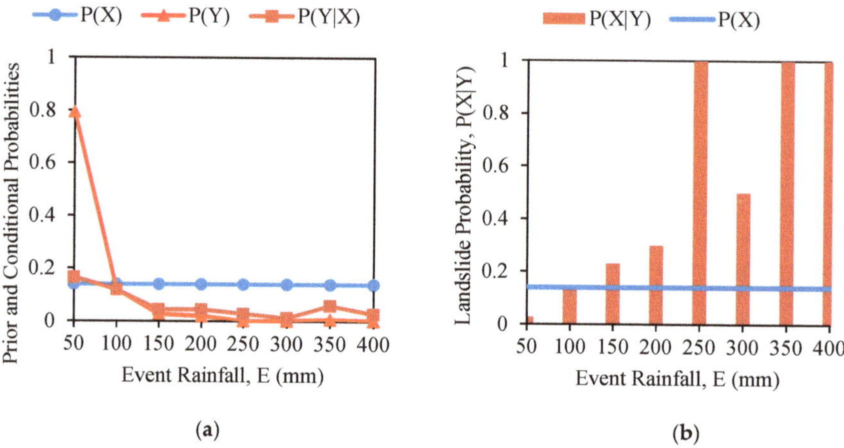

Figure 4. *Cont.*

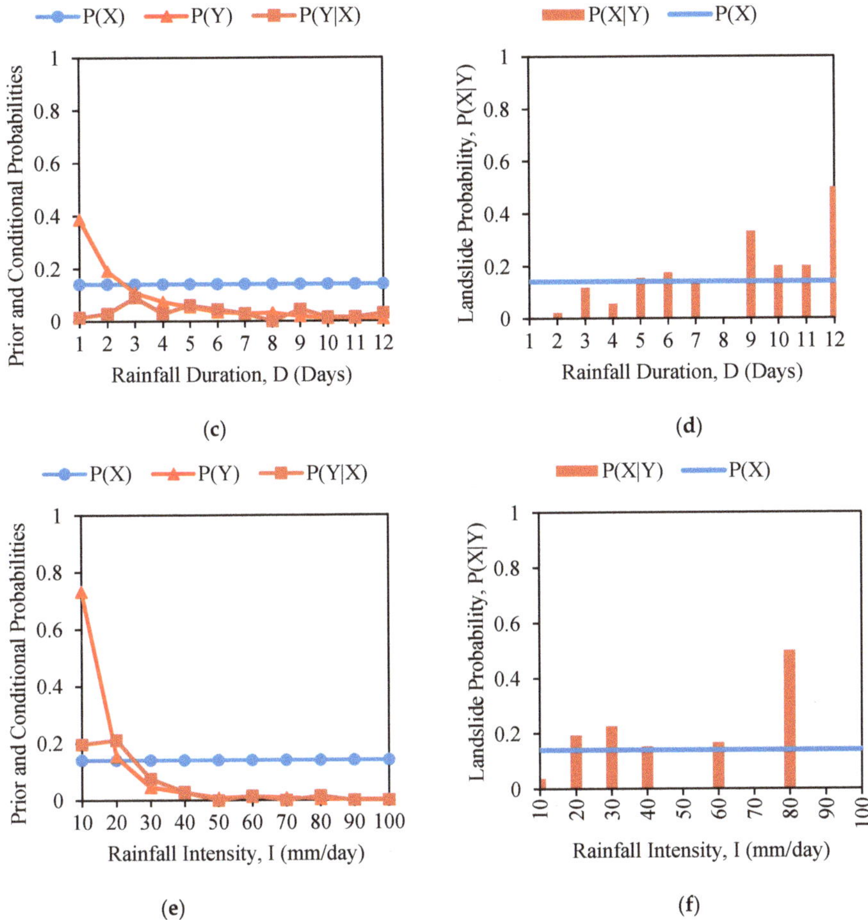

Figure 4. One-dimensional probability considering (**a**,**b**) Event Rainfall (**c**,**d**) Rainfall Duration (**e**,**f**) Rainfall Intensity for Chukha region.

The probability values are also hugely affected by the number of landslide events in each bin. There can be a huge difference in the values due to slight variation in the data points. For the present study, the number of landslide events for rainfall intensity in the bin 10–20 mm/day is 26, however, the number of points for 80–90 mm/day is only 5. This suggests that validation of the probabilities needs to be carried out and the results need to be enhanced with the addition of more data.

The result for two-dimensional Bayesian probability for landslide occurrences in Chukha has been depicted in Figure 5. The results indicate that maximum probability reaches 1 for the intensity of 30–40 mm/day with a duration of 7 days as well as the intensity of 40–50 mm/day with a duration of 6 days. However, the value of probability reaching an extreme amount can be attributed to low sample size and a slight variation in sample size can immensely affect the data. [18] determined probabilities for Kalimpong region which is roughly 100 km from the study area depicted that a precipitation event with intensity higher than 30 mm/day for 3 days represents the highest probability of 0.67. A sensitivity analysis of the calculated probabilities was performed to determine the variation of the results with a change in rainfall event parameters [19]. The input values of the parameters were varied and the changes in values of probabilities were ascertained. The variations include the definition of rainfall event definition and the period of analysis. The rainfall event for the probability determination was defined as the number of consecutive days of rainfall. For sensitivity analysis, we varied the minimum daily rainfall in intervals of 1mm and the corresponding probabilities were calculated which depicted similar trend as shown in Figures 4 and 5. The probabilities were recalculated by decreasing the analysis period and the changes in probabilities were in accordance to the number of events being considered. The results showed the increase in probability values for certain bins due to few numbers of landslide and rainfall events. The results show that the probabilities varied between 12%–22% for various bins of different rainfall parameters, however the peak values of landslide probability were in the same rainfall class. The analysis also suggested that one-dimensional results are more sensitive compared to two-dimensional Bayesian analysis.

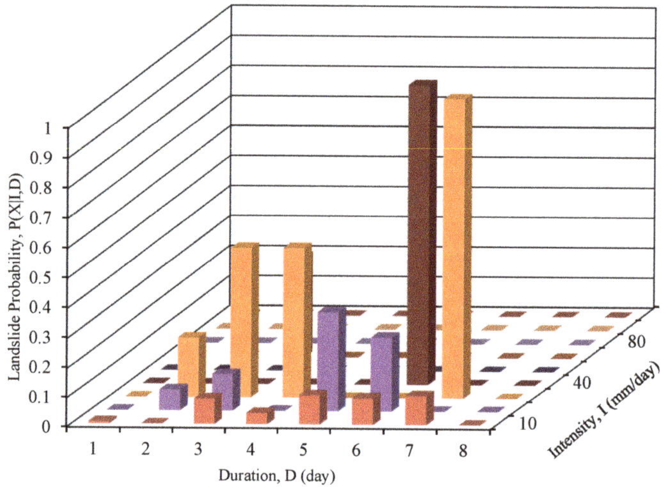

Figure 5. Two-dimensional probability considering rainfall intensity and rainfall duration parameters.

The validation of the probabilities was determined using the landslide record for 2015. During this period, there were 38 rainfall events and five landslide events with cumulative rainfall being 2698.83 mm (Figure 6).

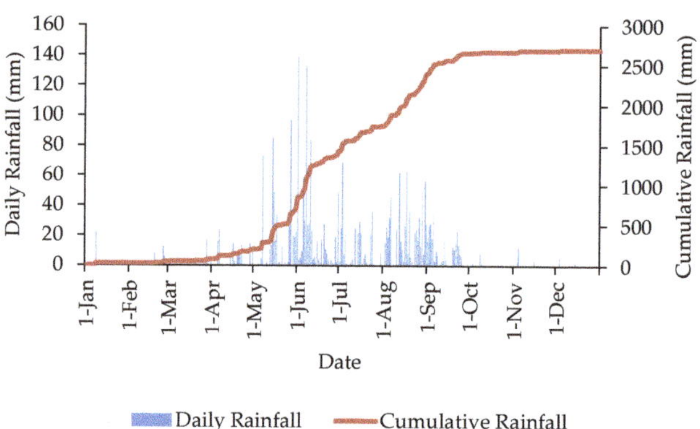

Figure 6. Daily and cumulative rainfall (in mm) of 2015.

Reference [13] defined regional based thresholds using an algorithm-based approach and determined that 69.7 mm of total rainfall for 24 h can cause landslides. Considering this value, the probability in terms of event rainfall is 0.14. To evaluate the performance of any model, threat score is considered as a useful metric [21]. Threat score is defined as:

$$TS = \frac{TP}{TP + FN + FP} \quad (6)$$

where TS is threat score, TP is true positive, FN and FP being false negatives and false positive respectively [25]. For the present study the various values of the Equation (6), TP = 4, FP = 1, FN = 2, which makes TS = 0.57. The result shows that the model has a good potential to be used for early warning system and can be improved with the further addition of data.

6. Conclusions

Landslides can be considered as one of the most devastating natural disasters due to their reoccurrence, economic damages and loss of human lives. Most landslides are shallow and caused due to rainfall which can be torrential downpours or slow and continuous. Therefore, it is imperative to understand the relation between rainfall and landslide incidences. Significant numbers of landslides are occurring throughout the Himalaya and as a part of it, Bhutan is not left out. The studies on understanding the relation of landslide incidences with rainfall have been rarely studied in this part of the world, so the present paper calculates the results of a probabilistic approach using Bayes' theorem using data between 2004–2014 to understand the effect of rainfall parameters for landslide initiation. The results were determined using a single along with a combination of rainfall parameters. The results were further validated using the landslide data of 2015 for the event rainfall parameter. The conclusions from the study are:

- The use of a probabilistic approach can be a better approach than empirical thresholds as the latter provides a single value of a specific rainfall parameter for landslide incidences.
- The use of two-dimensional probability for determining probabilities for landslide events is better as compared to one-dimensional as the latter depicts that a single rainfall parameter may not be a significant factor to trigger landslides.
- The validation of the thresholds for event rainfall parameter depicts that the model has an accuracy of 57%. However, with the addition of more landslide records and temporal rainfall data, the accuracy will improve. The use of such technique would help in setting up an operational early warning system and help in landslide mitigation.

Author Contributions: R.S. analyzed the data and wrote the article, K.D. collected the data and contributed in writing the article.

Funding: The research was funded from BRACE project (NERC/GCRF NE/P016219/1) granted to Raju Sarkar.

Acknowledgments: The authors are thankful to National Center for Hydrology and Meteorology, Royal Government of Bhutan for providing rainfall data and the Border Roads Organisation (Project DANTAK), Government of India for providing landslide data.

Conflicts of Interest: The authors declare no conflict of interest.

References

1. Froude, M.J.; Petley, D.N. Global fatal landslide occurrence from 2004 to 2016. *Nat. Hazards Earth Syst. Sci.* **2018**, *18*, 2161–2181. [CrossRef]
2. Wilson, R.C.; Wieczorek, G.F. Rainfall thresholds for the initiation of debris flows at La Honda, California, Environ. *Environ. Eng. Geosci.* **1995**, *1*, 11–12. [CrossRef]

3. Peruccacci, S.; Brunetti, M.T. TXT-tool 4.039-1.1: Definition and Use of Empirical Rainfall Thresholds for Possible Landslide Occurrence. In *Landslide Dynamics: ISDR-ICL Landslide Interactive Teaching Tools*; Sassa, K., Tiwari, B., Liu, K.F., McSaveney, M., Strom, A., Setiawan, H., Eds.; Springer: Berlin, Germany, 2018.
4. Guzzetti, F.; Peruccacci, S.; Rossi, M.; Stark, C.P. Rainfall thresholds for the initiation of landslides in central and southern Europe. *Meteorol. Atmos. Phys.* **2007**, *98*, 239–267. [CrossRef]
5. Brunetti, M.T.; Peruccacci, S.; Rossi, M.; Luciani, S.; Valigi, D.; Guzzetti, F. Rainfall thresholds for the possible occurrence of landslides in Italy. *Nat. Hazards Earth Syst. Sci.* **2010**, *10*, 447–458. [CrossRef]
6. Segoni, S.; Piciullo, L.; Gariano, S.L. A review of the recent literature on rainfall thresholds for landslide occurrence. *Landslides* **2018**, *15*, 1483–1501. [CrossRef]
7. Melillo, M.; Brunetti, M.T.; Peruccacci, S.; Gariano, S.L.; Roccati, A.; Guzzetti, F. A tool for the automatic calculation of rainfall thresholds for landslide occurrence. *Environ. Model. Softw.* **2018**, *105*, 230–243. [CrossRef]
8. Gariano, S.L.; Brunetti, M.T.; Iovine, G.; Melillo, M.; Peruccacci, S.; Terranova, O.; Vennari, C.; Guzzetti, F. Calibration and validation of rainfall thresholds for shallow landslide forecasting in Sicily, Southern Italy. *Geomorphology* **2015**, *228*, 653–665. [CrossRef]
9. Nikolopoulos, E.I.; Crema, S.; Marchi, L.; Marra, F.; Guzzetti, F.; Borga, M. Impact of uncertainty in rainfall estimation on the identification of rainfall thresholds for debris flow occurrence. *Geomorphology* **2014**, *221*, 286–297. [CrossRef]
10. Marra, F.; Destro, E.; Nikolopoulos, E.I.; Zoccatelli, D.; Creutin, J.D.; Guzzetti, F.; Borga, M. Impact of rainfall spatial aggregation on the identification of debris flow occurrence thresholds. *Hydrol. Earth Syst. Sci.* **2017**, *21*, 4525–4532. [CrossRef]
11. Peres, D.J.; Cancelliere, A.; Greco, R.; Bogaard, T.A. Influence of uncertain identification of triggering rainfall on the assessment of landslide early warning thresholds. *Nat. Hazards Earth Syst. Sci. Dis.* **2017**, *18*, 633–646. [CrossRef]
12. Staley, D.M.; Kean, J.W.; Cannon, S.H.; Schmidt, K.M.; Laber, J.L. Objective definition of rainfall intensity–duration thresholds for the initiation of post-fire debris flows in southern California. *Landslides* **2013**, *10*, 547–562. [CrossRef]
13. Gariano, S.L.; Sarkar, R.; Dikshit, A.; Dorji, K.; Brunetti, M.T.; Peruccacci, S.; Melillo, M. Automatic calculation of rainfall thresholds for landslide occurrence in Chukha Dzongkhag, Bhutan. *Bull. Eng. Geol. Environ.* **2018**. [CrossRef]
14. Berti, M.; Martina, M.L.V.; Franceschini, S.; Pignone, S.; Simoni, A.; Pizziolo, M. Probabilistic rainfall thresholds for landslide occurrence using a Bayesian approach. *J. Geophys. Res.* **2012**, *117*, F04006. [CrossRef]
15. Do, H.; Yin, K. Rainfall Threshold Analysis and Bayesian Probability Method for Landslide Initiation Based on Landslides and Rainfall Events in the Past. *Open J. Geol.* **2018**, *8*, 674–696. [CrossRef]
16. González, A.; Caetano, E. Probabilistic rainfall thresholds for landslide episodes in the Sierra Norte De Puebla, Mexico. *Nat. Res.* **2017**, *8*, 254–267. [CrossRef]
17. Dikshit, A.; Satyam, N. Rainfall Thresholds for Landslide Occurrence in Kalimpong Using Bayesian Approach. In Proceedings of the Indian Geotechnical Conference, Guwahati, India, 14–16 December 2017.
18. Dikshit, A.; Sarkar, R.; Satyam, N. Probabilistic approach toward Darjeeling Himalayas landslides—A case study. *Cogent Eng.* **2018**, *5*, 1537539. [CrossRef]
19. Dikshit, A.; Satyam, N. Probabilistic rainfall thresholds in Chibo, India: Estimation and validation using monitoring system. *J. Mt. Sci.* **2019**, *16*, 870–883. [CrossRef]
20. Chang, K.T.; Chiang, S.H.; Lei, F. Analysing the relationship between Typhoon-triggered landslides and critical rainfall conditions. *Earth Surf. Proc. Landf.* **2008**, *33*, 1261–1271. [CrossRef]
21. Marques, R.; Zêzere, J.; Trigo, R.; Gaspar, J.; Trigo, I. Rainfall patterns and critical values associated with landslides in Povoação County (São Miguel Island, Azores): Relationships with the North Atlantic Oscillation. *Hydrol. Proc.* **2008**, *22*, 478–494. [CrossRef]
22. Lagomarsino, D.; Segoni, S.; Rosi, A.; Rossi, G.; Battistini, A.; Catani, F.; Casagli, N. Quantitative comparison between two different methodologies to define rainfall thresholds for landslide forecasting. *Nat. Hazards Earth Syst. Sci.* **2015**, *15*, 2413–2423. [CrossRef]
23. Dunning, S.A.; Rosser, N.J.; Petley, D.N.; Massey, C.R. Formation and failure of the Tsatichhu landslide dam, Bhutan. *Landslides* **2006**, *3*, 107–113. [CrossRef]

24. Gansser, A. *Geology of the Bhutan Himalaya*; Birkhaüser Verlag: Basel, Switzerland, 1983; p. 181.
25. Corominas, J.; van Westen, C.; Frattini, P.; Cascini, L.; Malet, J.P.; Fotopoulou, S.; Catani, F.; Van Den Eeckhaut, M.; Mavrouli, O.; Agliardi, F.; et al. Recommendations for the quantitative analysis of landslide risk. *Bull. Eng. Geol. Environ.* **2014**, *73*, 209–263. [CrossRef]

MDPI
St. Alban-Anlage 66
4052 Basel
Switzerland
Tel. +41 61 683 77 34
Fax +41 61 302 89 18
www.mdpi.com

Hydrology Editorial Office
E-mail: hydrology@mdpi.com
www.mdpi.com/journal/hydrology

www.ingramcontent.com/pod-product-compliance
Lightning Source LLC
LaVergne TN
LVHW070636170726
843515LV00004B/264
* 9 7 8 3 0 3 6 5 2 1 7 7 0 *